Lecture Notes in Biomathematics

Managing Editor: S. Levin

60

Population Genetics in Forestry

Proceedings of the Meeting of the IUFRO Working Party
"Ecological and Population Genetics"
held in Göttingen, August 21–24, 1984

Edited by H.-R. Gregorius

Springer-Verlag
Berlin Heidelberg New York Tokyo

Editor

Hans-Rolf Gregorius
Abteilung Forstgenetik und Forstpflanzenzüchtung, Universität Göttingen
Büsgenweg 2, 3400 Göttingen, Federal Republic of Germany

Mathematics Subject Classification (1980): 92-02, 92-06, 92A10, 92A15, 92A17

ISBN 3-540-15980-0 Springer-Verlag Berlin Heidelberg New York Tokyo
ISBN 0-387-15980-0 Springer-Verlag New York Heidelberg Berlin Tokyo

Printing and binding: Beltz Offsetdruck, Hemsbach/Bergstr.
2146/3140-543210

Introduction

When we consider the main object of forestry, the tree, it immediately becomes clear why experimental population geneticists have been so hesitant in making this object a primary concern of their research. Trees are very long-living organisms with generation intervals frequently exceeding those of their investigators by multiples. They virtually exclude, therefore, application of the classical methods of population genetics since these are based on observing genetic structures over generations. This situation, where the limits set to observation are so severe, particularly requires close cooperation between theory and experiment. It also requires careful consideration of results obtained for organisms other than trees, in order to gain additional insights by comparing the results for trees with those for other organisms.

Yet, the greatest challenge to population and ecological genetics probably originates from the fact that forests are very likely to be the most complex ecosystems of all, even in some cases where they are subject to intense management. This complexity, which equally comprises biotic and abiotic factors varying both in time and space, makes extremely high demands on the adaptational capacity and thus flexibility of the carriers of such an ecosystem. Longevity combined with immobility during the vegetative phase, however, appears to contradict the obvious necessity of adaptational flexibility in forest tree populations when compared with short lived and/or mobile organisms. Mobility enables individuals to escape locally and temporarily unfavourable environmental conditions, and short generation intervals promote the quick response of a population's gene pool to changing environments. The biology of tree species seems to allow for neither of these advantages. Measured against their considerable evolutionary success, we must thus expect trees to have evolved highly sophisticated genetic systems of survival and reproduction, which possibly differ markedly from those of other organisms.

Understanding the principles of these systems is not just of academic interest, but it also is of vital importance for the benefit of man, as has become so drastically obvious in the last decades all over the world. For the time being we can only speculate on the extent of genetic impoverishment of forest tree species which has already been caused by the careless destruction of forests, environmental pollution and by poorly understood management, silvicultural and breeding strategies (by no means only in developing countries).

The present volume summarizes the papers contributed to a meeting of the working party Population and Ecological Genetics of the International Union of Forestry Research Organisations (IUFRO) held in Göttingen, August 1984. Being the first of its kind, this volume can only claim to give a necessarily incomplete impression of the status quo of research in forest population genetics.

It is striking that the majority of papers deals both experimentally and theoretically with mating systems in natural or artificially founded populations. Most attention is given to the degree of outcrossing, genetical techniques of estimating outcrossing (mainly based on enzyme markers), non-randomness of cross-fertilization, gene flow via pollen dispersal, genotypic structures in the seed production, and sytems of sexuality. Fertility selection, particularly if it occurs in a sexually asymmetric fashion, is apparently gaining increasing interest as a factor promoting genotypic diversity in combination with mating and sexual systems. The idea common to such studies consists in explaining the genetic basis for survival which is determined at the zygotic stage.

Another area of intense activity is genetic differentiation within and between populations of forest trees. The contributions to this field cover geographic and environmentally caused differentiation, effects of forest management and domestication on genetic diversity, effects of environmental pollution, and methods of measuring genetic differentiation. The practical implications of this research pertain to provenance selection, management of genetic resources, and hybrid breeding. From an evolutionary point of view the study of patterns of genetic differentiation form the basis for investigations on sympatric and allopatric speciation.

The financial support of the meeting by the Deutsche Forschungsgemeinschaft and the Niedersächsische Minister für Wissenschaft und Kunst is greatfully acknowledged. The generous help of G. Namkoong, H.H. Hattemer and F. Yeh in planning and arrangement made this meeting possible. The organisational work of B. Peter and E. Gillet and others laid the basis for the pleasant atmosphere. Special thanks go to the reviewers for their valuable comments and suggestions.

Göttingen, June 1985

Hans-Rolf Gregorius

TABLE OF CONTENTS

PART I :

TREE BREEDING

The Population Genetic Basis of Breeding Theory

Gene Namkoong

ABSTRACT

Genetic management of forest tree species requires a long-term perspective on the development of breeding populations. To achieve any recurrent selection objectives, we must understand gene interactions, the distinctions between mating demes and selection demes, and the stability of these interacting systems. The present understanding of gene actions in forest trees and the interaction of selection and mating systems are discussed, and some implications are drawn for breeding in partially managed forests.

INTRODUCTION

It seems obvious that population genetic theories have had remarkably little impact on breeding practices of domesticated plants and animals. Even for forest trees, which are more exposed to natural events than other crops, the study of natural selection and evolution has been a separate and often ignored field of inquiry. In spite of protests to the contrary that breeding populations contain large base populations, tree breeding programs generally concentrate on simple recurrent selection in single breeding populations that are susceptible to rapid reductions in effective population sizes. It is tempting to follow the examples of crop plants for which simple recurrent selection has produced genetic response for many generations. As long as some cursory attention is paid to effective population size, simple breeding methods can generally produce long-term cumulative gains in quantitatively inherited traits. Furthermore, as long as the time and technology are available, any problems with the advanced varieties can be solved by crossing or backcrossing into them, or substitute varieties can be made available by breeding from some relatively unimproved stock. For such breeding methods, the major influence of population genetic theories has been on organisms or populations of indirect interest. That is, population genetics is legitimately useful only for studying the structure of genetic variability in the wild relatives of the useful varieties or the effect that competitor species or unmanaged pest species may have.

In forestry, the initial results of selection and breeding have generally succeeded in producing substantial gains in economically important traits. However, the relatively long growth period from breeding to harvest and from breeding cycle to breeding cycle creates problems in evaluating gain and in defining what an optimal breeding program should achieve. In addition, the initial distribution of alleles in populations and populations over areas are largely unknown, and if they ever were in a stable equilibrium, have undoubtedly been substantially disturbed by human effects on the forest ecology. Furthermore, since forests will generally have to occupy extensive areas of lower economic productivity, per unit area, than agronomic crops, they are managed with lower intensity than farm lands. Since they are long-lived, their pests and pathogens have relatively shorter life cycles. This implies that forest trees will grow under less controlled environmental regimes, and hence will have to coexist with and be bred for production and coevolution with other species. Thus genetic management of forest tree species requires a long term perspective on the development of breeding populations and is more complicated than that of traditional agronomic species. To achieve any recurrent selection objectives requires us to understand gene interactions and to be efficient in land use. It also requires us to understand the distinctions that can be made between mating demes (subpopulations) and selection demes (multiple selection objectives). It also requires that we understand the stability of these interacting systems and the extent to which we have been and will be able to re- or de-stabilize them.

If these considerations are some of the major concerns of tree breeders which can be addressed by population geneticists, then this conference will well serve the audience of breeders if it deepens our understanding of how selection works, and the extent to which selection effects are confounded with the effects of other forces. In this paper, the present understanding of gene actions in forest trees, the interaction of selection and the mating system, and the implications for breeding in partially managed forest ecosystems will be discussed.

Gene Actions

One of the major breeding questions asked in forestry has been whether simple recurrent selection can be used as the primary breeding system instead of hybrid breeding methods. Ignoring pure line and pedigree breeding methods, the question for breeders focuses on whether a general heterozygote superiority or balancing selection exists and whether its effects can be captured at multiple loci. If present, are such effects significant for general vigor and vegetative productivity or limited to just certain traits? And if heterozygosity is a causal agent of general benefit, then the simpler breeding systems may fail to capture much of the gain potential achievable by finding the appropriate allelic mixtures for economically important traits. In fact, simple truncation selection and intercrossing could decrease yield. But if such effects are not important, then simple recurrent selection in breeding programs of sufficient size would accumulate alleles such that deleterious alleles are purged and partial dominance gene actions would allow cumulative gain to be realized. The contrast in viewpoint is evident in the juxtaposition of suggestions that a general heterozygosity is a very important element in vegetative growth and viability (Linhart et al., 1979), while most tree breeders practice some form of simple recurrent selection (Zobel and Talbert, 1984). While there is some ambiguity in the breeding programs with respect to the assumed genetic model, there are almost no programs that test and breed for reciprocal cross performance.

Unfortunately, the evidence from population genetic studies of forest trees is weak. Direct studies on the existence of deleterious mutants in trees as revealed by selfing and crossing studies do not support the hypothesis of heterosis. In general, estimates of the number of lethal equivalent alleles carried in conifer populations is high. However, if heterozygotes of these alleles are superior in fitness, then the frequency of A' should equilibrate at $q_e = s/(s + t)$ in a 2 allele system with fitnesses:

$$W_{AA} = 1-s$$

$$W_{AA'} = 1$$

$$W_{A'A'} = 1-t, \text{ where A' is the mutant.}$$

Heterosis implies that $s > 0$, and $t > 0$, and hence the closer to equivalence, the closer would q_e be to 0.5. Clearly, this is impossible for multiple loci in trees since cross pollination of mutant carriers fails to reveal any high frequency of lethals. On the other hand, if W_{AA} is close to $W_{AA'}$ and $W_{A'A'}$ is close to zero, then q_e is approximately $2\mu/s$ or $\sqrt{\mu/s}$. Thus, if $\mu \approx 10^{-6}$, and $s \approx 10^{-3}$, then q_e is in the range of 10^{-2} to 10^{-3}. In this range of low frequency, the allelic effects cannot be distinguished from a model of allele effects in which the mutants are maintained in the population simply by a selection-mutation balance against recessive lethals.

Similarly, in studies of recessive deleterious chlorophyll mutants, Snyder et al. (1965) find among 18 mutant carriers except for two neighboring trees that all mutant alleles were unique. Both of these results indicate low frequency of deleterious mutants; the latter by direct test, and the former by the high number of lethal equivalents revealed by selfing and the low mortality of crosses. Both therefore indicate that mutants are maintained only by recurrent mutation and hence that any observed inbreeding depression may well be the result of random mutations at a large number of loci carried at low frequency.

While such evidence is fairly strong on such alleles of drastic effect, other allelic interactions may be either neutral in net effect on evolution or may have some form of balancing selection such that their distribution reflects either migration and mating behavior or selective variations. Thus, other data useful for discerning the importance of gene actions and their selection significance are measures of genotypic frequencies in natural populations. In cases where heterozygote frequencies at isozyme loci exceed those expected from the average allele frequency functions of Hardy-Weinberg expectations, it is often inferred that heterozygote superiority is the causal agent. Such observations have been made by Tigerstedt (1973) and Mitton and Grant (1980), and have been supported by observations that older stands display stronger departures than younger stands (Yeh et al., 1983) and where older stands display a stronger positive relationship between heterozygosity and growth measures (Ledig et al., 1983). While these results are highly suggestive, none can exclude the possibility that these observations reflect random genotypic distributions in different populations which had different initial frequency distributions. The evidence is limited largely to older stands of post reproductive age, with unknown initial frequency distributions. Obviously, the mating system can initiate populations with wide departure from Hardy-Weinberg expected frequencies and the effects of selection mortality could as well be random as directional, and could result in excesses or deficiencies in any genotypic class depending on the initial conditions. Sexually asymmetric genotypic contributions to the zygotic pool and other forms of frequency-dependent selection can also lead to stable polymorphisms without heterosis at any locus. Furthermore, as Ginzburg (1983) derives, the existence of equilibrium populations with polymorphic loci will necessarily yield a measure of average heterozygote superiority regardless of the form of selection. That is, even assuming initial Hardy-Weinberg genotypic frequencies, it is a tautology to state that heterozygote superiority causes polymorphisms -- both are phenomena of whatever selective causes exist. Therefore, unless specific genotypic effects can be demonstrated as heterotic or populations can be measured as cohorts with their viabilities and fecundities estimated (cf. Christiansen and Frydenberg, 1976; and Clegg et al., 1978) would we be able to convincingly assert or assume the validity of the heterosis hypothesis.

As a consequence, tree breeders have actually derived very little information on gene actions from population genetic studies. There is no strong evidence that breeders could not simply cull any low-frequency deleterious alleles which may exist by ordinary breeding and selection, since these might be merely additive or dominant in their intralocus effects. Thus, ordinary quantitative genetic studies may provide information as reliable as tree breeders are liable to get; hence, studies on estimates of genetic variances of economically useful traits are as useful as any other available studies. In such studies, questions of net neutrality in evolution are not relevant as long as genetic variations exist and the relative sizes of additive, dominance, and epistatic effects can be estimated. In addition, studies on genetic variation in viability related traits, such as growth response to environmental variations, are clear indications of the potential for selection for fine-tuned adaptability (see e.g. Campbell, 1972; Fryer and Ledig, 1972; Namkoong and Conkle, 1976). These studies merely indicate the existence of genetic variation which is the present result of various selection and migratory histories and hence do not give information on the causal mechanisms. In fact, it is not proven that natural selection affects these genotypic distributions or that such variations may have a net evolutionary neutrality.

Mating System

Even if most of the allelic variation is either neutral or only held in populations by a mutation-selection balance, the existence of high levels of variation in quantitative and isoenzyme traits within and between populations (Hamrick, 1983) is undisputed. A factor operating in most forest populations of climax species is the subdivision of their range, and in non-climax species, is the fluctuation of their range and their mating patterns. Since selection in the mating system by itself can generate nonequilibrium genotypic frequency distributions (Ziehe and Gregorius, 1981), it is to be expected that any confounding of viability and mating selection can also generate substantial variations in genotypic distributions. Since Levene's (1953) recognition that the structure of mating versus selection demes affects the existence of polymorphisms, it has been clear that the study of selection effects over wide environmental variations cannot be satisfactory if rangewide average effects are assumed for the analysis. We have shown that local selection effects, coupled with limited, sexually dependent migration, can have very strong effects on the existence of polymorphisms and their stabilities (Gregorius and Namkoong, 1983; Namkoong and Gregorius, 1985). Thus, populational subdivisions may exist in either viability or reproduction, or both, and result in allele frequency distributions which do not reflect selective optima either locally or globally.

When populations are physically separated and may exist in different sets of stress conditions, genetic divergence may be rapid and relatively easy to measure as in several *Pinus ponderosa* stands (Linhart et al., 1979). However, divergence in contiguous stands also exists, which is likely to reflect more homogenizing effects of migratory gene flow. Nevertheless, divergence is measurable in P. taeda (Roberds and Conkle, 1984), in P. rigida (Fryer and Ledig, 1972) where there is segregation by the temperature optimum for photosynthetic efficiency, and in Pseudotsuga menziesii (Campbell, 1972) where there are growth response differences by elevation of parent trees.

Any such patterns of divergence or lack of it (Yeh, 1979) may not reflect any steady state behavior. Within stand divergence may actually reflect very recent contact between formerly disjunct stands. On the other hand, an observed lack of divergence in average allele frequency may reflect recent intermatings among formerly divergent sets of populations as in some P. taeda populations (Roberds and Conkle, 1984). In such unstable population structures, the presence or absence of ecologically related allele frequency divergence cannot be taken by itself as evidence for the relative strengths of selection on migration patterns (Lewontin, 1984). However, Roberds and Brotschol (in press) find that the likely cause of linkage disequilibrium is not epistatic selection, but is more likely a reflection of localized mating patterns which generate local allelic associations in seedling populations of Liriodendron tulipifera.

It is particularly revealing when patterns of stand structure are not consistent within the same species. For example, the lack of differentiation among stands of P. monticola in northern Idaho is quite different from the observations of high levels of divergence among stands found in more southern latitudes which existed below the limits of glaciation (Rehfeldt, 1983). Similar differences in divergence patterns above and below the southern borders of glaciation in Pseudotsuga menziesii is interpreted by Critchfield (1984) as reflecting the recent migration of populations into the formerly glaciated northern areas from limited refugia. Since northern stands of this species contain high levels of genetic variation within stands (Yeh, 1979), the populations may soon resegregate into divergent populations. On the other hand, the reproductive system may be so variable as to allow different modes of evolution to exist in the different parts of a species range. While slow to respond to selective pressures, it is conceivable that secondary selection causes the variation.

It is thus strongly suggested that the history of migratory-mating patterns may have had a strong effect on allelic and genotypic distributions. Since the effects of selection are strongly conditioned by the mating system (Hedrick et al., 1978), and if selection is itself an effective force, then the evolutionary effects of all these forces are confounded. There is little evidence to support any hypotheses of broad, uniform types of selection at many loci in forest trees. Selection of some balancing form would have to be invoked, but too little is known of how local variations may affect that balance. Not only may the local balance be strongly reflective of the mating system, it may also be shifting both temporally and spatially. Thus, for the tree breeder to discern whether genetic variation in local adaptations exist and can be utilized in breeding programs, we need to know much more about both the mating and the selection system.

If there is a lesson to be learned from our attempts to apply population genetic research to breeding, it is that we require more information on the environment and the reproductive system. We need information on the extent of sexual differences in viability, fertility and mating success that are genetically determined. We need to know if outcrossing is genetically conditioned and if outcrossing is of a form that ties the whole species together or if, as in shrub and herb species, there is a strong potential for independent evolution among subpopulations. In particular for trees, we need to know if temporal variations and age related effects on selection and reproduction reduce the effective population size for any given outcrossing rate.

Selection in Forests

In addition to questions confounded with migration and mating, the nature of selection is not clear for organisms that have several life stages and which compete and interact with their own and other species and which do so in temporally and spatially variable environments. There are obvious genetic variations in response to environments and artificial selection can obviously change important responses; however, it is not clear which are the trade-offs and which the risks, nor how the individual tree or population responds to variations. Obviously, trees are perennial sessile organisms that achieve a size and longevity at the upper extremes of that achieved by earthbound organisms. Distinctively for trees, the secondary meristem and leaves provide most of the defense and maintenance systems while the primary meristem, beneath its protective dome, provides the apical growth around the crown and produces the reproductive organs. Genes that affect both meristems or their products clearly have complicated effects on vegetative and reproductive success which are likely to vary as the tree ages. Thus, even single locus effects would generally have to be considered to have pleiotropic effects which probably are dependent on the environment for their net effects on fitness. The measurement of such effects, and distinguishing the effects of countervailing selection forces between simple environmental effects versus effects in which feedback mechanisms may operate, and age-, density-, or frequency-dependent effects, is a central concern of population geneticists.

For forest tree breeders, however, it is necessary but not sufficient to understand evolution in order to successfully and efficiently create breeding populations. The necessity derives from the need to use the genetic variations, which evolutionary forces have provided, in more or less controlled selective breeding and culture treatments. However, the breeder also needs to consider optimal future designs for populations which may or may not mimic natural evolution. It would certainly behoove the breeder to understand niche theory since the evolution of gene actions creates the kinds of genotype-environment joint effects available; hence, understanding how genotypes perceive and react to environmental variations determines how genotypes are allocated among sites. This would be an obvious way to begin an optimum assignment of breeding populations to Target Populations of Environments (Comstock, 1977) or to consider how to develop optimal Multiple Index Selection Strategies (Namkoong, 1976) for multiple population breeding (Namkoong et al., 1980).

The high degree of uncertainty we have with respect to environmental variations is underscored by recent observations of pollution caused forest ecological crises in Europe (Waldsterben) and North America. Adaptability of forests to unanticipated environmental crises may be an important breeding objective which can even override economic yield objectives. It is therefore important for the forest tree breeder to consider how best to structure individual genotypes, stands, and their source breeding populations, to maximize adaptability to unanticipated crises.

Similarly, it behooves the breeder to understand the development of competitor and host/pathogen coevolutionary systems in order to at least understand the status of any interacting system. Furthermore, by understanding how genetic variations in density- and frequency-dependent effects can alter the ecology of the interaction, and how the ecology can alter the stability properties of genetic systems (Selgrade and Namkoong, 1984; Namkoong and Selgrade, in press), the breeder-forester may then understand how fluctuations in numbers and allele frequencies can follow determined paths. From such information, sources of alleles for breeding might be identified and natural systems of control studied for suggestions on how to construct stable systems. The breeder must then determine how sets of populations can be managed for an optimal gene-ecological system.

In forestry, with its more extensive management system and less intensive input of mechanical or chemical interventions, it is to be expected that the system for which the breeder must create population sets will be closer to the natural ecology than that of traditional crop plants. There would then be a greater necessity for tree breeders to understand the dynamics of the evolutionary history of, at least, the recent past, and that may be sufficient.

Literature Cited

Campbell, R. K. 1972. Genetic variability in juvenile height-growth of Douglas-fir. Silvae Genet. 21: 126-129.

Christiansen, F. B. and O. Frydenberg. 1976. Selection component analysis of natural polymorphisms using mother-offspring samples of successive cohorts. p 277-302. In: Population Genetics and Ecology, S. Karlin and E. Nevo, eds. Academic Press, New York. 832 pp.

Clegg, M. T., A. L. Kahler and R. W. Allard. 1978. Estimation of life cycle components of selection in an experimental plant population. Genetics 89: 765-792.

Comstock, R. E. 1977. Quantitative genetics and design of breeding programs. p 705-718. In: Proc. International Conf. on Quantitative Genetics. Aug. 16-21, 1976. Iowa State Univ. Press, Ames, IA. 872 pp.

Critchfield, W. B. 1984. Impact of the pleistocene on the genetic structure of North American conifers. p 70-118. In: Proc. Eighth North American For. Biol. Workshop. R. M. Lanner, ed. Utah State Univ., Logan, UT. 196 pp.

Fryer, J. H. and F. T. Ledig. 1972. Microevolution of the photosynthetic temperature optimum in relation to the elevational complex gradient. Can. J. Bot. 50: 1231-1235.

Ginzburg, L. R. 1983. Theory of Natural Selection and Population Growth. Benjamin/Cummings Publ., Menlo Park, CA. 160 pp.

Gregorius, H. -R. and G. Namkoong. 1983. Conditions for protected polymorphisms in subdivided plant populations. 1. Uniform pollen dispersal. Theor. Popul. Biol. 24: 252-267.

Hamrick, J. L. 1983. The distribution of genetic variation within and among plant populations. p 500-508. In: Genetics and Conservation. C. M. Schonewald-Cox, S. M. Chambers, B. MacBryde and L. Thomas, eds. Benjamin/Cummings Publ., Menlo Park, CA. 722 pp.

Hedrick, P., S. Jain and L. Holden. 1978. Multilocus systems in evolution. Evol. Biol. 11: 101-184.

Ledig. F. T., R. P. Guries and B. A. Bonefield. 1983. The relation of growth to heterozygosity in pitch pine. Evolution 37: 1227-1238.

Levene, H. 1953. Genetic equilibrium when more than one ecological niche is available. Am. Nat. 87: 331-333.

Lewontin, R. 1984. Detecting population differences. Am. Nat. 123: 115-124.

Linhart, Y. B., J. B. Mitton, K. B. Sturgeon and M. L. Davis. 1979. An analysis of genetic architecture in populations of ponderosa pine. pp 53-59. In: M. T. Conkle (tech. coord.), Proc. Symp. on Isozymes of North American Forest Trees and Forest Insects. USDA For. Serv., Gen. Tech. Rep. PSW-48. Pacific Southwest For. Exp. Stn.; Berkeley, CA.

Mitton, J. B. and M. C. Grant. 1980. Observations on the ecology and evolution of quaking aspen, _Populus tremuloides_, in the Colorado Front

Range. Am. J. Bot. 67: 202-209.

Namkoong, G. 1976. A multiple index selection strategy. Silvae Genet. 25: 199-201.

Namkoong, G., R. D. Barnes and J. Burley. 1980. A Philosophy of Breeding Strategy for Tropical Forest Trees. Tropical Forestry Papers No. 16, Oxford Univ., England. 67 pp.

Namkoong, G. and M. T. Conkle. 1976. Time trends in genetic control of ponderosa pine. For. Sci. 22: 2-12.

Namkoong, G. and H. -R. Gregorius. 1985. Conditions for protected polymorphism in subdivided plant populations. 2. Seed versus pollen migration. Am. Nat. 125: 68-84.

Namkoong, G. and J. F. Selgrade. [In press.] Frequency-dependent selection in logistic growth models. Theor. Popul. Biol.

Rehfeldt, J. 1984. Microevolution of conifers in the Northern Rocky Mountains: A view from common gardens. p 132-146. In: Proc. Eighth North American For. Biol. Workshop. R. M. Lanner, ed. Utah State Univ., Logan, UT. 196 pp.

Roberds, J. H. and J. V. Brotschol. [In press.] Linkage disequilibrium among allozyme loci in natural populations of _Liriodendron tulipifera_ L. Silvae Genet.

Roberds, J. H. and M. T. Conkle. 1984. Genetic structure in loblolly pine stands: allozyme variation in parents and progeny. For. Sci. 30: 319-329.

Selgrade, J. F. and G. Namkoong. 1984. Dynamical behavior of differential equation models of frequency and density dependent populations. J. Math. Biol. 19: 133-146.

Snyder, E. B., A. E. Squillace and J. M. Hamaker. 1965. Pigment inheritance in slash pine seedlings. pp. 77-85. In: Proc. Eighth Southern Conf. on Forest Tree Improvement, Savannah, GA. June 16-17.

Tigerstedt, P. M. A. 1973. Studies on isozyme variation in marginal and central populations of _Picea abies_. Herediatas 75: 47-60.

Yeh, F. C. 1979. Analyses of gene diversity in some species of conifers. p 48-52. In: M. T. Conkle (tech. coord.), Proc. Symp on Isozymes of North American Forest Trees and Forest Insects. USDA For. Serv., Gen. Tech. Rep. PSW-48. Pacific Southwest For. Exp. Stn., Berkeley, CA.

Yeh, F. C., A. Brune, W. M. Cheliak and D. C. Chipman. 1983. Mating systems of _Eucaplyptus citriodora_ in a seed production area. Can. J. of For. Res. 13: 1051-1055.

Ziehe, M. and H. -R. Gregorius. 1981. Deviations of genotypic structures from Hardy-Weinberg proportions under random mating and differential selection between sexes. Genetics 98: 215-230.

Zobel, B. and J. T. Talbert. 1984. Applied Forest Tree Improvement. John Wiley & Sons, New York. 505 pp.

HYBRIDISATION AND CYTOGENETICS OF EUROPEAN BIRCHES

I.R. Brown and D.A. Williams

SUMMARY

The complex cytology of European birches is considered from three points of view namely hybridisation in controlled and natural conditions, somatic chromosomal variation and the role of the dwarf birch in speciation and variation of arborescent birch.

European studies, including recent work at Aberdeen, have indicated that it is possible to produce viable seed and hybrid progeny from all combinations of *B. pubescens*, *B. pendula* and *B. nana* crosses. The success of hybridisation is subject to environmental influences and cytological factors such as relative ploidy levels of forests.

Although fertility of hybrids is low and there is no evidence of their ability to survive under natural conditions studies of Scottish birch populations appear to indicate that hybridisation and backcrossing contribute to the variation within and between populations. The situation is complicated by the occurrence of aneuploidy and somatic chromosome variation.

INTRODUCTION

The situation regarding crossing, hybridisation and cytogenetics in *Betula* is complex and there are important implications concerning tree breeding, populution genetics and ecology in birch and in any other species that may demonstrate similar cytological phenomena. In order to assess these implications it is necessary to draw together three of the strands of research that have been developed at Aberdeen University Department of Forestry since about 1969. The first concerns hybridisation under both controlled and natural conditions, the second has to do with the phenomenon of somatic chromosome variation and the possible existence of naturally occurring aneuploids and the third line of research, which has received least emphasis, is centred upon the role of

B. nana L., the dwarf birch, in speciation and variation in the arborescent European birches.

According to the Flora Europaea (Walters, 1964) there are two species of tree birches and one, *B. pubescens* Ehrh., is divided into three subspecies that are sometimes given specific rank. Linnaeus designated *B. pendula* Roth. and *B. pubescens* as a species complex to which he assigned the epithet *B. alba* L. Associated with these species at high altitudes and latitudes is *B. nana*. Both *B. pendula* and *B. nana* have 2n = 28 chromosomes while *B. pubescens* has 2n = 56 and the basic number for the genus is believed by many workers to be x = 7 rather than x = 14.

HYBRIDISATION

European literature abounds in references to hybrids between *B. pendula* and *B. pubescens* and in Britain, in nearly every area where birches grow, so called 'hybrids' have been recorded. In theory all putative hybrids can be checked by determining if they have the expected chromosome number of 2n = 42. In an extensive review of the subject Vaarama (1969) threw doubt on the authenticity of all but two out of more than 20 recorded hybrids some of which appeared to be *B. pendula* autotriploids. In an investigation of birches of intermediate morphology Jurkevic and Gubanov (1969) showed that they were all 2n = 56. Other evidence such as no, or limited, overlap in flowering times (Jentys Szaferowa, 1938) has also been presented to throw doubt upon the common occurrence of hybrids.

On the other hand all combinations of controlled crosses between *B. pubescens*, *B. pendula* and *B. nana* have produced viable seed and hybrid progeny (Eifler, 1958, 1960; Hagman, 1971; Williams, 1981; and Stern, 1963). Although naturally fruitful, birch is often difficult to manipulate under greenhouse and other artificial conditions necessary for controlled crosses. Thus crosses may fail at first attempt while in the following year viable seeds are produced in profusion. Hagman (1971) suggested that external conditions can affect the incompatability mechanism and Williams (1981) found that crosses involving 2n = 28 cytotypes were successful in years when spring temperatures were relatively low. Success with other crosses varied for no apparent reason. It has also been found (Clausen, 1970) that crosses in which the higher ploidy species is used as male parent are usually more successful than the reciprocal cross. The existence of such inconsistencies in results means that generalised published statements about the difficulty or otherwise of the hybrid cross that are based upon either small numbers of trees or on one year's crossing attempts cannot be taken at face value.

OCCURRENCE OF NATURAL HYBRIDS

Given this facility for intercrossing and results by Aston (1975) which showed that, in Scotland at least, there is sufficient overlap in flowering times for natural hybridisation to occur, the question arises as to why there is a relative paucity of authenticated recordings of the occurrence of tree birch hybrids in the wild. In the past, it has been assumed that any hybrids between *B. pubescens* and *B. pendula* would be intermediate in appearance but, given that the triploid hybrid would contain twice as many *B. pubescens* chromosomes as those from *B. pendula*, it is reasonable to suppose that gene dosage effects would cause the hybrid to look more like the former than the latter species. Thus in the field it is highly likely that any existing hybrid trees would be overlooked.

As part of a small tree improvement programme for *B. pendula* currently in progress at Aberdeen about 100 mother trees were selected and from these approximately 2,000 open pollinated progenies were raised in a plastic greenhouse (polytunnel). All the mother trees were fairly typical *B. pendula* with pointed leaves, double toothing, black patterned white bark, etc but amongst their progenies were 94 atypical seedlings originating from 23 mother trees growing on 13 sites. All the other seedlings in the polytunnel had, after some months continuous growth, developed verrucose main shoots and double-toothed, acutely-pointed leaves but the atypical trees had softly hairy shoots and hairy, bluntly-pointed leaves. Determinations of somatic chromosome numbers were made on expanding leaf bases and were approximately $2n = 42$. More detailed analysis of morphology was made after the seedlings had grown for three seasons in the open and developed short-shoot leaves which are usually assumed to be less variable than juvenile or long-shoot leaves. A step-wide discriminate analysis showed that morphologically they overlapped the range of both *B. pendula* and *B. pubescens* but tended to resemble the latter rather more than the former (Kennedy and Brown, 1983). They did so not only in leaf form but also in other traits such as rootability of cuttings. This contrasts somewhat with the findings of Eifler (1958) who described her hybrid seedlings as having pronounced heart-shaped bases (to such an extent that the basal lobes overlapped) and crinkled leaf edges. She compared them with the leaves of tetraploid *B. pendula* that had been produced by colchicine treatment and stated that they were very similar in appearance.

As has been reported previously by Stern (1963) the hybrids are highly sterile but, compared with Stern's findings, the hybrids produced by the Scottish *B. pendula* population do not show signs of hybrid weakness - at least under conditions of cultivation. Other hybrid seedlings could, of course, have died between sowing and transplanting without being recognised. In the progeny test the hybrids flowered precociously and some profusely but no viable seed have yet been produced from wind pollination. On one occasion, using a hybrid as pollen parent and a *B. pendula* growing in a polytunnel as the mother tree, three seedlings were produced two of which still survive. Their chromosome complements have not been determined and on casual inspection their leaves appear to resemble *B. pendula* rather than *B. pubescens*. These seedlings were originally rather weak but now, in their second year, they have become vigorous in contrast to their early struggle to survive.

BACK-CROSSING

If hybrids are triploid in constitution then it would be expected than on backcrossing to *B. pendula* and *B. pubescens* that 2n = 35 and 2n = 49 progenies would result. Stern (1963), however, found that the hybrid used as a pollen parent and crossed with *B. pendula* produced triploid progenies and in a similar cross with *B. pubescens*, tetraploid and pentaploid seedlings resulted. Thus he concluded that progenies were derived from unreduced pollen and egg cells or that there was a sharp discrimination against aneuploid plants. The production of unreduced gametes in *B. pendula* was suggested by Hagman (1971) to explain the production of progenies in both his and Eifler's hybrid F_1's that had chromosome complements of 2n = 56 and he further postulated that 50% of *B. pubescens* gametes were n = 14 to explain the high proportion of 2n = 28 progenies amongst hybrid F_1's. Studies in meiosis by Brown and Dawoody (1979) showed that such suggestions were supported by the existence of monads, dyads and octads in the microspore mother cells. While there is supporting evidence for the production of unreduced gametes (not least the occurrence of autotriploid *B. pendula*), there is also evidence for the existence of aneuploid birches which will be dealth with below.

MEIOSIS IN HYBRIDS

In late August 1982 efforts were made to study meiosis in the male catkins of the Scottish hybrids. Anthers were smaller than normal and hard to handle and, in addition, great difficulties were found in staining procedures. Eventually only four trees out of twenty produced analysable meiotic figures and even here, results must be regarded as very tentative. On another occasion (Brown and Dawoody, 1979) difficulties in staining and examination of chromosomes were associated with a high level of abnormality in those cells which could be examined. In this case too, difficulties in cell preparation may also indicate a high level of chromosomal and meiotic abnormality. Of the four trees studied, one had cells at metaphase I with either 14 or 28 bivalents, another had cells with 21 or 28 bivalents while in the third tree only cells with 21 bivalents were formed. Concerning the fourth tree, meiosis appeared to be normal but chromosome figures were unanalysable.

NON-HYBRID TRIPLOIDS

A longstanding and unsolved problem is presented by the existence of birch with 2n = 42 chromosomes that are neither autotriploids of *B. pendula* nor hybrids between that species and *B. pubescens* (Brown and Williams, 1982). They are fertile, produce seedlings with 42 chromomsomes on wind pollination, intercross with both *B. pendula* and *B. pubescens* and are indistinguishable on the basis of leaf morphology from *B. pubescens* (Brown and Dawoody, 1977, 1979 and Gardiner and Pearce, 1979). The chromosome numbers of these trees, determined on tissue from the bases of expanding leaves are more variable than those of the other two cytotypes. Counts from F_1 progenies are similarly variable but those resulting from crosses with *B. pendula* are approximately 2n = 35 and with *B. pubescens*, around 2n = 49. The trees were found as adults with ages ranging from approximately 35-70 years during a survey of two birch woods in N.E. Scotland that contained *B. pendula* in the lower reaches and *B. pubescens* ssp. *odorata* on the higher more exposed sites. Of a total of 174 trees 75 were 2n = 28, 84 were 2n = 56 and 15 were intermediate in number. Eleven of the latter were 2n = 42 and four originally classified as mixoploids had counts ranging from 2n = 45 to 51. Many of the intermediates were found in a boundary area of the main wood where there was a change from a preponderance of one of the main cytotypes to the other.

Various lines of evidence such as the results of leaf-isozyme analysis (Williams, 1981) indicate that the trees with intermediate chromosomes are aneuploid *B. pubescens*. If, as Hagman suggests, doubly reduced gametes are common in this species then an obvious hypothetical origin for the trees would be the union of non- and doubly-reduced *B. pubescens* gametes. Although the nearest *B. nana* are about 15 kilometres distant it is possible that the intermediates are the result of historic hybridisation, although this is in no way indicated by their external morphology.

CYTOTYPE SURVEYS

The work discussed above was eventually followed up by a more extensive survey of Scottish birch woodlands. Five areas were sampled, usually by line transects, to determine chromosome numbers of adult trees and their open-polinated progenies. The five sites were (1) a pure wood of *B. pubescens* ssp. *odorata* in N.E. Scotland near the original woodlands that produced the aneuploids, (2) what was originally thought to be a pure wood of *B. pendula* in central Scotland but subsequently a few *B. pubescens* were found on the upper margin, (3) a mixed *B. pendula* and *B. pubescens* ssp. *pubescens* stand in eastern Scotland and two areas (4) and (5) where *B. pubescens* ssp. *odorata* grew side by side with *B. nana*, area (4) in the far north and (5) in the Grampians. The first two woods consisted of trees with modal values of $2n = 56$ and $2n = 28$ respectively and all seedlings had the parental somatic values. Site (5) was remote from other woodlands and had rather sparsely distributed *B. pubescens* which all yielded the expected count of $2n = 56$. In none of these sites nor in woodlands of pure *B. pubescens* in N. Scotland remote from *B. pendula* sampled in earlier work (Williams, 1981) were any aneuploid birches found. Gill and Davy (1983) investigated birch woodland in eastern England where there were many morphologically intermediate forms. Using squashes of shoot apical meristems they determined the chromosome numbers of 50 trees and their seedlings and found no evidence of triploid trees nor of aneuploidy. Earlier surveys in Scotland (Berrie, 1953) and northern England (Croker, 1955) also failed to reveal the existence of aneuploid trees. Thus it would appear that the aneuploid trees reported by Brown and Dawoody (1977) represent a unique or very rare case.

GLEN PROSEN

Studies on the trees of site (3) were related to the occurrence of hybrid seedlings and did not concern the aneuploids. This wood in Glen Prosen in N.E. Scotland was originally selected for its trees of outstanding phenotype to be used in the breeding programme mentioned above. Originally the trees were assumed to be fairly typical and there were no more than the usual difficulties of assigning trees to their appropriate species. But since 60% of all the hybrid seedlings originated from eight trees growing here the area was thought worthy of further study (Brown *et al*, 1982).

All counts on birch leaf bases done at Aberdeen have demonstrated a greater or lesser degree of variation in chromosome number from cell to cell within leaves. Some of this variation is accounted for by error which increases as numbers of chromosomes per cell rise but the residual variation is real and is lowest in cytotypes with a mean around 28, highest in those with mean counts of 35 and counts of trees with about 42 or 56 chromosomes are intermediate with more variation in the 42's than 56's. Attempts to use root tips have usually failed due to lack of active cell division.

In Glen Prosen then, it turned out that amongst the *pendula*-like trees all about 40 years old only three could be judged to be typical *B. pendula* since the others, although having counts between 2n = 27 and 34, were associated with very high coefficients of variation. Seedlings originating from wind pollination of these trees had counts varying from 2n = 26 to 44 and even the 'normal' *B. pendula* produced seedlings with high chromosome numbers. Concerning the adult *pubescens*-type trees also of similar age two had complements of 2n = 37 and the rest had counts well within the expected range for *B. pubescens*. Seedlings selected at random were very variable and counts ranged from means of 2n = 26 and 56. Some of the low count seedlings from *B. pubescens* mother trees were indistinguishable from *B. pendula* seedlings. In this original investigation rather small numbers of cells were counted for the seedlings so the trees were re-examined in a subsequent survey some years later. On the whole the previous results were confirmed although on this occasion the progenies from the *pendula*-type trees contained very few abnormalities (Figure 1), and individual trees varied between assessments. Variation between seedlings was mirrored by variation within plants; examples are shown in Table 1. Compared with the range of mean counts obtained for *B. pubescens* trees growing in the north of Scotland beyond the natural range of *B. pendula*, the Glen Prosen progenies show an extraordinary range of counts (Figure 1).

It is quite possible that some of the adult trees in Glen Prosen are themselves of hybrid origin and it was originally the sort of disturbed site that is traditionally associated with the survival of hybrid plants.

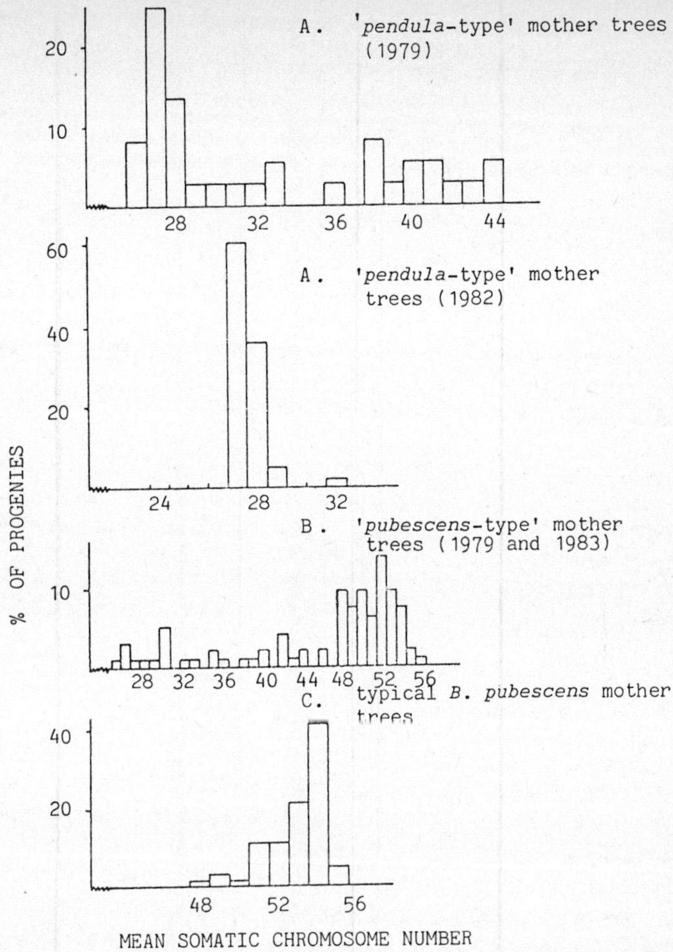

Figure 1 Mean somatic chromosome counts of wind polinated seedlings from A. '*pendula*-' and B. '*pubescens*-type' trees in Glen Prosen and from C. typical *B. pubescens* trees in the N. of Scotland.

On the other hand their fertility is much higher than any recorded for artificial hybrids although compared with other sites the proportion of filled seed at 19% and germination around 11% (Brown *et al*, 1982) is low but was only measured on one occassion.

Attempts were made to study meiosis in the Glen Prosen trees but the material was more than usually difficult to stain and prepare for examination. The trees that might have proven most interesting to examine, number 10 with a somatic number of 2n = 37 and number 22 which produced seedlings of very low chromosome number (Table 1) could not be analysed at all. On the whole meiosis appeared to be essentially regular but as

SEED TREE No.	PROGENY No.	Mitotic counts	2n ± S.D.	No. of Cells counted	COEF VAR
1	1	25, 27, 28, 26, 26 28, 28, 28, 26, 28	27.0 ± 1.2	10	4.3
(1982) (2n = 26.9 ± 1.7)	2	28, 28, 28, 27, 28 27, 28, 27, 24, 27	27.2 ± 1.2	10	4.5
	3	28, 27, 26, 28, 26 28, 28, 30, 28, 30	27.9 ± 1.4	10	4.9
10	1	48, 49, 54, 40, 53 54, 54, 56, 54, 56	51.8 ± 4.9	10	9.5
(1982) (2n = 37.3 12.7)	2	51, 42, 55, 56, 53 54, 45, 53, 52, 51	51.2 ± 4.4	10	8.6
	3	56, 52, 56, 49, 49 49, 56	52.43 ± 3.5	7	6.7
	4	48, 48, 49, 47, 50 52, 51, 49, 52, 51	49.7 ± 1.8	10	3.6
	5	45, 54, 50, 53, 56 51, 53, 54, 56, 47	49.9 ± 7.0	10	14.0
	1		35.4 ± 6.6	21	18.7
(1979)	2		30.0 ± 6.0	5	20.0
	3		24.9 ± 3.7	9	15.0
	4		55.7 ± 1.5	3	2.8
22	1	30, 42, 28, 32, 27 45, 42, 56, 56, 42	40.0 ± 10.7	10	26.7
(1982) (2n = 37.6)	2	52, 48, 50, 53, 56 56, 40, 54, 40, 52	50.1 ± 5.9	10	11.7
	3	47, 46, 41, 51, 48 50, 51, 50, 56, 42	48.2 ± 4.5	10	9.3
	4	28, 46, 30, 52, 28 42, 32, 30, 50, 29	36.7 ± 9.7	10	26.5
	5	46, 38, 50, 35, 48 35, 39, 42, 33, 40	40.6 ± 5.8	10	14.3

Table 1   Somatic chromosome complements of the progeny of *pubescens*-type seed trees in Glen Prosen numbers 10 and 22. Tree number 1, a *B. pendula*, is included for comparison.

was found with the aneuploid *P. pubescens* clumping of chromosomes was common and a variety of cytotypes produced empty pollen mother cells as well as monads, dyads and tetrads. Such phenomena may reflect unseen failures and disturbances of meiosis.

Given the production of unreduced, doubly reduced and normal gametes from the species types and gametes of unknown constitution from trees that might be of hybrid origin there is a large number of permutations of combinations of chromosomes in the seedlings. Given the autoploidy of *B. pubescens* and the possible inclusion of *B. pendula* chromosomes in its genome then even pure *B. pendula* seedlings are a possible product of the Glen Prosen *pubescens*-type trees. Further resolution of this problem does not, it seems, lie in continued study of chromosomes but perhaps in a chemotaxonomic approach.

B. NANA

Finally there is the question of the involvement of the dwarf birch, *B. nana*, in the morphology if not the differentiation of *B. pubescens* into sub-species. The idea that introgression occurs between *B. nana* and *B. pubescens* gained currency with the results of Elkington's studies (1968) in Iceland. Based upon a comparison of hybrid indices of *B. nana* in Iceland and Scotland, *B. pubescens* in Iceland and *B. pubescens* in England it was concluded that introgressive hybridisation accounted for the fact that the Icelandic *B. pubescens* was of intermediate hybrid index between the dwarf birch and the English *B. pubescens*. Previously Löve and Löve (1956) had pointed out that the triploid hybrid would be sterile and they suggested that the morphological variation in *B. pubescens* populations of Iceland derived from sources other than introgressive hybridisation. Vaarama and Valanne (1973), on the basis of morphological and physiological similarities between *B. pubescens* ssp. *tortuosa* and *B. nana*, concluded that introgressive hybridisation was a principal factor in the evolution of the *B. tortuosa* species/sub-species. They also reported that a *B. nana* x *B. pubescens* hybrid has produced seed from which progeny had been raised and had survived in cultivation. In a later publication (Sulkinoja and Valanne, 1980) it was reported that viable progenies from triploid hybrids varied from 2n = 42 to 2n = 70 and some aneuploid numbers also occurred. Seed germination was very low, 0 to 8.7%, and pollen meiosis irregular with many aborted grains.

It was against this background of somewhat conflicting and uncertain information that area (4) on Ben Loyal in N. Scotland was chosen for study. Kenworth *et al* (1972) established that triploid birches found here were hybrids between *B. nana* and *B. pubescens* and it was decided to carry out mitotic and meiotic studies on the *B. pubescens* in the hope that any irregularities discovered might indicate the occurrence of introgressive hybridisation. In the event no such evidence was found since meiosis was regular and chromosome counts of the open-pollinated progenies were regular. The are is very remote and due to difficulties in timing collection of leaves in the proper condition no somatic counts were established for the adult *B. pubescens*. On one visit to the site some low growing birches no more than 1.5 m high were seen to be fruiting (grazing pressure here is high and usually only inaccessible or hidden shrubs flower and fruit), seeds were collected and trees raised. The results were quite unexpected since some of the progenies very closely resembled *B. nana* and had chromosome couts of 2n = 28, some were intermediate between *B. nana* and *B. pubescens* and were 2n = 35, while others from the same mother tree appeared to be normal *B. pubescens*. This segregation of morphological types would suggest the occurrence of backcrossing but neither the source of the pollen nor the cytotype of the seed tree are know.

CONCLUSIONS

Studies of chromosomes and birch cytotypes have produced results whose complexity parallel that of the morphological variation. The degree of somatic variation is puzzling and may as in other species be an indication of hybridisation. On the other hand gametes may show less variability. This could explain why trees such as number 10 in Glen Prosen (Table 1), although having very variable somatic counts produced, in 1982, quite unexceptional progeny but in 1979 there was a greater degree of variability. Differences between years, however, could be accounted for by differences in the constitution of the pollen cloud. If the adult trees are of hybrid origin it is difficult to reconcile their relatively high fertility with that reported for artificial and natural F_1 hybrids. It has already been shown (Brown and Williams, 1982) that when the distribution of individual cell counts for a population is examined, peaks in frequency occur at all multiples of seven between 14 and 56. Is it possible for somatic tissues to change with age so that previously infertile trees produce tissues that contained balanced sets of chromosomes to allow a reasonably high proportion of regular meiosis to occur?

Fertility in aneuploids can be explained on the assumptions of seven being the basic chromosome number and that, although now fully diploidised, the 2n = 28, 42 and 56 cytotypes may act, at times, as tetraploids, hexaploids and octaploids. Thus an aneuploid *B. pubescens* is one that has lost a maternal and paternal set of seven chromosomes. Duplication of sets of chromosomes that may have changed through mutation but still retain a degree of homology would explain the presence of multivalents in all three cytotypes. Hybrid sterility can be accounted for by the hypothesis that *B. pubescens* and *B. pendula* share only certain seven-chromosome sets.

Such speculation is not very profitable until methods can be devised to identify and characterise sets of chromosomes or by use of techniques such as DNA extraction and renaturation. This had already been tried at Aberdeen but the precipitated DNA would not spool up on first attempts, presumably due to the DNA being broken into too short lengths. This seems to be a method of analysis that is worth following up.

Another point to be made is that trees live a long time and birch produces vast amounts of pollen and seed and thus no matter how low fertility of hybrids may be their participation in the evolution of the modern birch and its varieties could be very important. At first sight it appeared that to propose hybridisation as the answer to the problems of the origin of morphological and physiological diversity in birch was taking the easy way out. Now, although the evidence is neither conclusive nor plentiful, it appears very likely that the European arborescent birches are very complex products of hybridisation involving extant and perhaps extinct species.

REFERENCES

ASTON, D. The taxonomy and genecology of birch. Ph.D. thesis, University of Aberdeen (1975).

BERRIE, A.M. A study of the Scottish birch with special reference to the genetics and ecology of the species. Ph.D. thesis, University of Glasgow (1953).

BROWN, I.R. and AL-DAWOODY, D.M. Cytotype diversity in a population of *Betula alba* L. New Phytol. *79* : 441-453 (1977).

BROWN, I.R. and AL-DAWOODY, D.M. Observations on meiosis in three cytotypes of *Betula alba* L. New Phytol. *83* : 801-811 (1979).

BROWN, I.R., KENNEDY, D. and WILLIAMS, D.A. The occurrence of natural hybrids between *Betula pendula* Roth. and *B. pubescens* Ehrh. Watsonia *14* : 133-145 (1982).

BROWN, I.R. and WILLIAMS, D.A. Proceedings of a symposium 'Birch', Sept. 1982, Edinburgh. Royal Soc. Edin. (in press).

CLAUSEN, K. Interspecific crossability tests in *Betula*. Proceedings IUFRO Section 22 meeting on Sexual Reproduction of Forest Trees, Varparanto, Finland (1970).

CROKER, B. The autecology of *Betula* species in the vicinity of Sheffield. Ph.D. thesis, University of Sheffield (1955).

EIFLER, I. Kreuzungen zwischen *Betula verrucosa* und *Betula pubescens*. Züchter *28* : 331-336 (1958).

EIFLER, I. Untersuchungen zur individuellen Bedingtheit des Kreuzungserfolges zwischen *Betula pendula* und *Betula pubescens*. Sivae Genetica *9* : 159-165 (1960).

ELKINGTON, T.T. Introgressive hybridisation between *Betula nana* L. and *B. pubescens* Ehrh. in north-west Iceland. New Phytol. *67* : 109-118 (1968).

GARDINER, A.S. and PEARCE, N.J. Leaf shape as an indicator of introgression between *Betula pendula* and *B. pubescens*. Trans. Bot. Soc. Edinburgh *43* : 91-103 (1979).

GILL, J.A. and DAVY, A.J. Variation and polyploidy within lowland populations of the *Betula pendula/B. pubescens* complex. New Phytol. *94* : 433-451 (1983).

HAGMAN, M. On self- and cross-incompatibility shown by *Betula verrucosa* Ehrh. and *Betula pubescens* Ehrh. Comm. Inst. Forest. Fenn. *73*(6) : 1-125 (1971).

JENTYS-SZAFEROWA, J. Biometrical studies in the collective species *Betula alba* L. 2. The possibility of hybridisation between species *Betula verrucosa* Ehrh. and *B. pubescens* Ehrh. Int. Bod. Los.

Ponstw. Roszpr. i. Sprow S.A. 40 Worszawa (1938).

JURKEVIC, I.D. and CUBANOV, K.D. Chromosome numbers in some birch forms. Dokl. Acad. Nauk., USSR *13* : 635 (1969).

KENNEDY, D. and BROWN, I.R. The morphology of the hybrid *Betula pendula* Roth. x *B. pubescens* Ehrh. Watsonia *14* : 329-336 (1983).

KENWORTHY, J.B., ASTON, D. and BUCKNALL, S.A. A study of hybrids between *Betula pubescens* Ehrh. and *B. nana* L. from Sutherland - an integrated approach. Trans. Bot. Soc. Edinb. *42* : 517-539 (1972).

LOVE, A. and LOVE, D. Cytotaxonomical conspectus of the Icelandic Flora. Acta Horti. Gothob. *20* : 65-290 (1956).

STERN, K. Uber einige Kreuzungsversuche zur Frage des Vorkommens von Arthybriden *Betula verrucosa* x *B. pubescens*. Deutsche Baumschule, Aachen *15* : 1-10 (1963).

SULKINOJA, M. and VALANNE, T. Polyembryony and abnormal germination in *Betula pubescens* subsp. *tortuosa*. Rep. Kevo Subarctic Res. Stat. *16* : 31-37 (1980).

VAARAMA, A. Induced mutations and polyploidy in birch, *Betula* species. Final Report, Part 1, Department of Botany, The University of Turku, Finland (1969).

VAARAMA, A. and VALANNE, T. On the taxonomy, biology and origin of *Betula tortuosa* Ledeb. Rep. Kevo Subarctic Res. Stat. *10* : 70-84.

WALTERS, S.M. *Betula* in : Flora Europaea, Vol. 1, 57-58. Cambridge University Press (1964).

WILLIAMS, D.A. Chemotaxonomy and cytology of birches in Scotland. Ph.D. thesis, University of Aberdeen.

PART II :

MATING SYSTEMS

MATING SYSTEM ESTIMATION IN FOREST TREES: MODELS, METHODS AND MEANINGS

A.H.D. Brown, S.C.H. Barrett and G.F. Moran

ABSTRACT

Isozyme markers have stimulated efforts to measure the mating system in natural and planted populations of forest trees. When open-pollinated progenies have been surveyed for allozyme variation, the arrays have usually been analyzed in terms of the model of mixed self-fertilization and random outcrossing. This procedure has encountered several problems of which the more difficult are due to heterogeneity in outcrossing rates or in pollen allele frequencies. Recent models which deal with these problems and enrich the study of plant mating systems include multilocus estimation, measurement of 'effective' selfing and of differential male fertility. More reliable estimates of outcrossing, which these methods encourage, will be needed to predict selection gains, management practices and optimal conservation strategies in forest trees.

INTRODUCTION

One topic attracting increased attention among forest geneticists is the analysis of mating systems operating in various populations; be they natural stands, plantations or seed orchards. That mating systems are crucial in determining population genetic structure, and the character of genetic transmission to subsequent generations has long been recognised (e.g. as reviewed by Stern and Roche, 1974). Yet two factors have recently emerged to stimulate this interest further. These are the growth of theoretical work on the evolution of mating systems, and second, the availability of isozyme markers with which to follow mating events after the fact. This paper is concerned with the problem of measuring mating systems. The focus will be on the use of the mixed mating model, primarily in its application to temperate, hermaphroditic forest species. Our aim is to consider the problems encountered in its application, and some of the remedies devised. Finally we outline questions which estimation experiments might address and the importance of the results obtained. More general discussions of estimation are given by Ritland (1983), and citations therein.

MATING SYSTEM OF MIXED SELF-FERTILIZATION AND RANDOM OUTCROSSING

Deviation from purely random mating in plant populations has most commonly been specified by a model assuming a mixture of two types of gametic union. In this model each zygote is assumed to result from either a self-fertilization, with fixed probability s, or as fertilization with a pollen grain chosen at random from the whole population with probability t=1-s. This model was first applied to predominantly self-pollinated crops (Jones, 1916; Fyfe and Bailey, 1951), and later used widely in agricultural and natural populations of both inbreeding and outbreeding species including herbaceous plants and forest trees.

In such a model, estimation of the genetic consequences of the mating system is based on the joint behaviour of genotypes at one or more marker loci through the cycle of one mating event. Since codominant isozyme loci have become prevalent as genetic markers, the treatment will be based on such genes. The basic data are the segregation patterns of the frequency of progeny genotypes derived from known or inferred maternal genotypes. In the latter case, progeny size is assumed to be sufficient so that the maternal genotype is reliably known. The simplest transitions can be specified for a single diallelic locus, with alleles A_1 and A_2 as in Table I. The observed number of progeny genotypes for each maternal genotype is coded as (O_i; i=1, ...,6). The expectations are based on assuming the mating system of mixed selfing and random outcrossing, where the allele frequency of A_1 in the pollen is p. In this simplest (diallelic) case, the expected frequency of A_1 A_2 progeny from A_1A_2 maternal plants is independent of the mating parameters (s,p) and is omitted here for convenience. However this class of progeny must be added to generalize the scheme to multiple alleles (Cheliak et al, 1983), multiple independent loci (Ritland and Jain, 1981), and for other mixed mating models (such as mixed outcrossing, selfing and apomixis (Marshall and Brown, 1974)).

In gymnosperms, all the progeny of heterozygous maternal plants can yield information on the mating system provided both the maternally derived haploid megagametophyte is assayed with each progeny (Shaw and Allard, 1982). Complete gametic classification is also theoretically possible for loci expressed in the triploid endosperm of angiosperms such as for seed storage protein genes. In these special cases, classes 3 and 4 in Table 1 are redefined such that O_3 includes the endosperm-progeny combinations (A_1; A_1A_1) and (A_2; A_1A_2) whereas O_4 includes the combinations (A_1; A_1A_2) and (A_2; A_2A_2). The expectations remain as defined in Table 1.

TABLE 1 Basic maternal/offspring matrix for estimation of the parameters of the mating system of mixed self-fertilization and random outcrossing.

	Genotypes		Frequencies	
	Maternal	Progeny	Observed	Expected
1.	A_1A_1	A_1A_1	O_1	$N_{11}(1-X)$
2.		A_1A_2	O_2	$N_{11}\ X$
3.	A_1A_2	A_1A_1	O_3	$N_{12}(1-X+Y)/2$
4.		A_2A_2	O_4	$N_{12}(1+X-Y)/2$
5.	A_2A_2	A_1A_2	O_5	$N_{22}\ Y$
6.		A_2A_2	O_6	$N_{22}(1-Y)$

where

$$X = tq$$
$$Y = tp$$
$$N_{11} = O_1 + O_2$$
$$N_{12} = O_3 + O_4$$
$$N_{22} = O_5 + O_6$$

As each maternal genotype provides one degree of freedom in the progeny array, at least two maternal types are required to estimate both mating parameters. When progenies from all three maternal genotypes are available, a test for goodness-of-fit is possible. Brown et at (1975) detail the maximum likelihood estimation procedure for this case, and Ritland (1983) provides a more general formulation.

ASSUMPTIONS OF THE BASIC MIXED MATING MODEL

The formulation of the basic model in Table 1 is designed to show its features and assumptions. The formal statistical and genetic assumptions required for estimation can be summarized as follows -

1. Within each maternal parent, the progeny genotypic classes are independent, identically distributed, multinomial random variables.
2. The values of both mating system parameters t and p (and therefore their one-to-one transformations X and Y) are uniform over maternal plants.
3. Segregation of the alleles in heterozygous maternal plants is strictly Mendelian in a 1:1 ratio for both pollen and ovule production.
4. Selection does not occur between fertilization and the assay of progeny genotypes.

These fundamental assumptions are sufficient to specify the construction of the model and the estimation of its parameters. However the assumptions themselves, or the interpretation of the estimates may require several other biological features. In particular, the model assumes inbreeding arises only through self-fertilization. This implies that the pollen involved in each outcrossing event is a random gamete sampled from the entire population. Each outcross is strictly an independent sample from the same uniform population of pollen. The genotype of each outcrossing pollen grain is independent of the maternal genotype, and of that of other outcrossing pollen grains included in the same maternal family. In a similar vein, the probability of outcrossing is assumed to be constant for all maternal plants, unaffected by maternal genotype, and independent of whether any other seed in the sample for assay is an outcross or a self.

PROBLEMS WITH SINGLE-LOCUS ESTIMATES OF OUTCROSSING RATE

As the above model and estimation procedure has been widely applied, a number of problems have come to light. Here we list and discuss several of these.

1. Estimates of outcrossing exceeding unity

When the method is applied to populations of predominantly outcrossing species, estimates of t exceeding unity have occasionally been obtained. Such so-called "biologically unreasonable" values have perturbed several workers. This problem does not arise from Wahlund effects, or necessarily from invalidity of the mixed mating model. Two primary causes are (1) sampling effects within the context of a valid mixed mating model; and (2) disassortative mating.

The first cause can be illustrated with a simple example. Suppose the true parametric value of t is unity and the population strictly adheres to the assumptions of the model. Consider a marker locus with equally frequent alleles. The expected values of X and Y are both 0.5. Consider a simple set of data based on only homozygous maternal plants (A_1A_1 and A_2A_2). On half the occasions such an assay is made, the expected result would be that O_2 would exceed $N_{11}/2$ and hence the estimate of X would exceed 0.5. Likewise on half the occasions (independent of the behaviour of O_2), the estimate of Y would exceed 0.5. Therefore in one half of such experiments the estimate of outcrossing (t = X+Y) would exceed unity, just by chance alone, despite the complete adherence of the population to the model.

In attaching a biological interpretation to such estimates, it seems natural to truncate the estimate of t at t=1.0. There are however two important dangers with this procedure. First, when a whole series of estimates based on different loci, populations or species are to be summarized as a mean of estimates, the use of truncated values will bias the overall average estimate of outcrossing downwards. Second, if truncation at t = 1.0 is an automatic step of the estimation procedure, such as with the E.M. algorithm (see below), the iteration procedure may yield a biased estimate of the pollen allele frequency (p).

A clear example of the effects of disassortative mating on estimates of outcrossing is that based on the polymorphism governing style-length in tristylous, self compatible *Eichhornia paniculata* (Barrett and Brown, unpublished data). Open-pollinated progenies from maternal plants of known style-length were classified, and estimates of outcrossing based on the mixed mating model were found to exceed unity (Table 2). However when the progeny expectations were adjusted to allow outcrosses only from legitimate pollinations (which amounts to disassortative matings for style-length), values of t less than unity were obtained.

2. Heterogeneous estimates from different marker loci

When the same progeny arrays, and hence the same mating events, have been assayed at several marker loci, it is frequently found that the segregations of the separate loci, yield markedly different estimates of outcrossing. This was a common finding for estimates based on morphological marker loci (Harding and Tucker, 1964), and was usually attributed to selection acting differentially on the marker loci. With isozyme markers this explanation may be less attractive. A possible contributing factor to this problem may be that sampling variances of single-locus estimates based on the Cramer-Rao inequality are underestimates of these variances. Further, when the number of maternal plants is restricted, sampling variance among loci may be increased, because each marker locus may test overlapping sets of mating events and with unequal precision. Thus heterogeneity of estimates over loci may not necessarily argue that single loci estimation, or the mixed mating model is unsatisfactory.

TABLE 2. Estimates of outcrossing in tristylous Eichhornia paniculata

Style Morph	Mixed Mating Model	Disassortative Mating Model
Long	1.39	0.90
Mid	1.49	0.93
Short	1.18	0.84

3. Heterogeneity of outcrossing rate - temporal, spatial, population density and maternal genotype

The assumption of uniformity of outcrossing rate in a population must be regarded as unrealistic. Indeed several studies have provided evidence of heterogeneity related to the above four parameters (Hamrick, 1982). The detection of such heterogeneity is obviously important if seed harvests of the desired level of outbreeding are needed. Further, factors influencing variation in outcrossing are of prime importance in understanding the evolution of mating systems.

In reality the experimenter may have to contend with variation in outcrossing arising from more than one source. For example, in a study of a temporal sequence of three seed crops in Eucalyptus delegatensis, Moran and Brown (1980) found that outcrossing estimate was highest in the oldest crop. The difference could merely be year-to-year variation in outcrossing. Alternatively, greater loss of inbred seed in the oldest crop than in the youngest crop, might be the source of the heterogeneity. In an attempt to resolve this difference Moran and Bell (unpublished data) estimated outcrossing in eight crops from the same 33 trees of Eucalyptus stellulata collected over three years. The single-locus estimates are given in Table 3, grouped by year of mating, and show both sources of heterogeneity could be present. Temporal variation was apparent in that the outcrossing rate in 1980 was above average. As selfing was higher in 1979, there was opportunity for detecting elevated estimates of outcrossing when the seed formed in that year was allowed to remain on the tree.

TABLE 3. Temporal variation in outcrossing rate in one population of *Eucalyptus stellulata* (Moran and Bell, unpublished)

Year Fertilized	Year Assayed	Locus: Aph	Adh	Pgi	Sdh	Average
1981	+1	0.59	0.98	0.81	0.75	0.78
1980	+1	0.93	0.84	0.91	0.87	0.89
	+2	0.81	0.88	0.84	0.87	0.85
1979	+1	0.65	0.76	0.76	0.67	0.71
	+2	0.73	0.81	0.86	0.83	0.81
	+3	0.87	0.75	0.95	0.66	0.81
1978	+2	0.73	0.72	0.81	0.62	0.72
	+3	0.69	1.04	0.76	0.64	0.78
Total		0.75	0.84	0.84	0.75	0.80

Less attention has been paid to whether the routine estimate of outcrossing based on a model assuming uniformity, is a biased estimate of the population mean, when this assumption is false. A simple example illustrates that this could be a problem. Consider a population with only A_1A_1 and A_2A_2 maternal plants in the ratio p:q. If the outcrossing rates are t_1 and t_2 respectively, then the expected value for the outcrossing estimate of a model assuming uniform outcrossing is $t_1q + t_2p$. The actual frequency of outcrossed seed is $t_1p + t_2q$, so that the bias in the estimate is $(t_1-t_2)(q-p)$. This calculation indicates that marker loci with near equal allele frequencies are to be preferred, as likely to be more robust to variation in outcrossing.

4. Heterogeneity of pollen allele frequencies among and within maternal genotypes

Another problem arises when the outcross pollen differs among maternal plants in allele frequency. This problem was soon encountered when the methods originally developed for agricultural populations were applied to natural populations. Genetic differentiation of subpopulations on a microgeographic scale, coupled with restricted pollen dispersal leads to a partial correlation between pollen genotype and maternal genotype. Estimates of outcrossing overlooking this correlation will be biased downwards (Ennos and Clegg, 1982). Alternatively, disassortative mating may be viewed as another breakdown in this assumption in which outcrossing estimates would be inflated (as in the *Eichhornia* example). Many of the developments to be discussed below (multilocus procedures, "effective selfing") address the problem of subpopulation structure.

It is difficult to tell whether heterogeneity arises from variation in outcrossing rate, or from variation in pollen frequencies. For example, a chi-square contingency test may be used to check for uniformity of progeny frequencies among the individual maternal plants of the same genotype. When this test detects heterogeneity, for frequency of progeny genotypes, it is not apparent whether outcrossing rates differ, or pollen allele frequencies differ (or both).

5. Multiple alleles

The occurrence of multiple alleles at a locus although providing more statistical information for mating system parameters, leads directly to numerical complications in computing the estimate. Programs and procedures are available to use the added information. Alternatively the researcher may lump the rarer alleles and reduce the dimensions of the task. No rules have been established to guide such lumping. It can be shown that for random samples from populations in inbreeding equilibrium, the pollen allele frequencies enter the expression for the theoretical variance for the maximum likelihood estimate of t, solely as powers of the product pq. Computation of the variance for a fixed value of t, and various values of p, shows that the variance is a minimum for p=q=0.5. Therefore lumping of alleles should proceed to yield the two-allele case with synthetic allele frequencies nearest to 0.5. However, for more complex models which invoke a greater number of parameters, multiallelic data may be needed. (Ritland, 1984.)

6. Segregation distortion

In heterozygous maternal plants, meiotic drive at micro - or mega sporogenesis or gametophytic competition thereafter could seriously distort the 1:1 expected gametic production, and hence affect genotypic frequencies in the progeny array. Yet the arrays of maternal heterozygotes generally yield limited statistical information in the estimates of the mating system parameters. Therefore biases to estimates arising from this source are likely to be minor. However the model can be modified to include the estimation of such effects if this is deemed necessary.

In angiosperms, distortion of either ovule or pollen segregation (or both) will affect the expectations in Table 1. In particular, the expected frequency of A_1A_2 progeny on A_1A_2 females will either exceed or decline from 1/2 and this may be used as a chi-square test for distortion (Clegg, 1983). In gymosperms, when the relevant megagametophytic data are available, distortion for ovule frequencies would not affect the estimates of X and Y, as is seen in the more complex model below.

Let us assume that the ratio of $A_1:A_2$ ovules matured on A_1A_2 maternal plants is k:1-k, and the ratio of $A_1:A_2$ self-fertilizing pollen grains is ℓ: 1-ℓ. In angiosperms, the observed and expected numbers of the three progeny genotypes are shown in Table 4A. With the extra degree of freedom (four altogether), the four parameters (k, ℓ, X and Y) can be estimated. The maximum likelihood estimators are obtained by equating observed with expected frequencies (Bailey, 1951). The four independent equations are provided by classes 2 and 5 in Table 1, and 3' and 4' in Table 4A. More restricted models are possible by substituting ℓ=1/2 for segregation distortion only for ovules, or k=ℓ for equal distortion in both gametes.

For gymnosperms, there is an extra class and hence an extra degree of freedom. The expectations are given in Table 4B, and it is clear that the maximum likelihood estimator of k is

$$k = (O_{31} + O_{41}) / N$$

The M.L. estimates of ℓ, X and Y would have to be obtained by numerical procedures. As noted above, this also implies that the departure of k from 1/2 in the gymnosperm case would not affect the estimation of X and Y if it were based on the basic mixed mating model.

MORE COMPLEX PROCEDURES AND MODELS

Along with the growth in attempts at measuring outcrossing rates in plant populations have emerged several procedures to handle the problems they encountered.

TABLE 4 Expected numbers of progeny genotypes in maternal heterozygotes when segregation distortion is present.

A. Angiosperms

Class	Progeny	Observed Number	Expected Number
3'	A_1A_1	O_3	$N[k\ell s + ky]$
	A_1A_2	$N-N_{12}$	$N[(1-k)(\ell s+Y) + k(s(1-\ell)+X)]$
4'	A_2A_2	O_4	$N[(1-k)(1-\ell)s + (1-k)X]$
Total		N	N

B. Gymnosperms

Mega -gametophyte genotype	Embryo genotype	Observed Number	Expected Number
A_1	A_1A_1	O_{31}	$N[k\ell s + kY]$
A_1	A_1A_2	O_{41}	$N[k(1-\ell)s + kX]$
A_2	A_1A_2	O_{32}	$N[l(1-k)s + (1-k)Y]$
A_2	A_2A_2	O_{42}	$N[(1-k)(1-\ell)s + (1-k)X]$
Total		N	N

1. The E-M algorithm

Recently Cheliak et al., (1983) have used the Expectation-Maximization (EM) algorithm for obtaining the maximum likelihood estimates of mating system parameters. The major advantage of this procedure is its facility in coping with a high number of alleles at the marker loci. In principle it can be extended to handle multiple loci. As formulated, it yields estimates of outcrossing bounded strictly between 0 and 1. In any given set of data, it yields identical estimates for outcrossing with the strictly comparable maximum likelihood analysis (M.L) provided the latter is less than unity. However, in the case cited above, where M.L. gives the estimate of t exceeding 1.0; the value from the algorithm is t=1.0. Caution is thus needed in interpreting E.M. estimates of t=1.0 because the simultaneous E.M. estimates of pollen allele frequencies will be biased. The E.M. procedure does not provide values for the variance of the estimates, except in the case of a single parameter.

2. Multilocus procedures

Several of the above problems (notably 1,2,4,5 and 6) have led to the suggestion that a single estimate of outcrossing based on the joint behavior of several marker loci may be more robust than separate single-locus estimates (Shaw et al., 1981). In addition, the formulation of standard errors for the average estimate of outcrossing strictly requires a multilocus approach, rather than by combining the standard errors of single-locus estimates. This is because the scoring of an increased number of loci per individual only increases the probability of detecting an outcrossing event. The precision of the estimate of the number of such events is limited by the total number of zygotes sampled.

This leads us to the concept of detection probability. Each progeny of a maternal array can be classified as to whether it is a genetically marked outcross or not. The probability that an outcross will be detected depends on whether the pollen grain carries a non-maternal allele. When only a small number of marker loci are employed, this probability may vary considerably over maternal genotypes. Green et al. (1980) used a procedure allowing different detection probabilities for each maternal genotype in a three-locus analysis of outcrossing in *Lupinus albus*. The main problem with this method is that computing the detection probabilities requires making assumptions about the frequencies of pollen multilocus genotypes.

Shaw et al. (1981) developed and simplified this approach further to cope with many loci and hence multilocus maternal genotypes. Their estimator essentially assumes that the same detection probability applies to all outcrosses irrespective of maternal genotype. Although this assumption is strictly incorrect, the error introduced is negligible when the number of loci is large, so that the detection probability is high. They investigated the effect of gametic disequilibrium on their estimator. Correlation among marker loci in effect lowers the detection probability of outcrosses. If estimation procedes ignoring the correlation then the estimates of outcrossing are biased downwards. The simplest means of avoiding this problem would be to test for the presence of disequilibrium, and use in the mating analysis, the scores of only one locus from any correlated pair. With these caveats in mind, the method of Shaw et al., although approximate, and statistically inefficient, should prove particularly useful in obtaining estimates from data on multiple loci.

Ritland and Jain (1981) present a complete M.L. estimation procedure based on many independent loci. They demonstrated numerically that the multilocus estimate is less affected by selection and non-random outcrossing than are single locus estimates. Data from a large number of loci would require considerable computer storage and iteration time. However, their analysis indicates that in most circumstances three to four loci would approach the minimum variance possible.

3. Subpopulation differentiation

A significant complicating factor in the analysis of natural populations, as distinct from agricultural populations or plantations, is the occurrence in the former of subpopulation structure. In natural plant populations, genetically alike individuals occur together because of restricted pollen dispersal or seed migration, and because of localized selection. Estimates of outcrossing based on the simple mixed mating model will usually be biased downwards. Some contrived types of subpopulation structure, such as over-dispersed plantations of genotypes can lead to upward biases in estimates (Ellstrand and Foster 1983). Such mating designs are relevant for seed orchards attempting to maximize disassortative mating.

Recently Ritland (1984) has developed a model aimed at estimating the unbiased "effective selfing" rate. The model derives estimates of selfing rates of inbred parents as opposed to outbred parents. The concept of "effective selfing" is designed to include both true selfing and crossing with related individuals. The estimation procedure includes maternal genotypic frequencies, apparently assumes equilibrium and yields higher standard errors because of the contribution of parental sampling.

4. Outcrosses sharing paternity

Schoen and Clegg (1984) have analyzed another departure from the assumptions of the mixed model. Suppose there is correlation between the pollen genotypes of the outcross progeny within a single progeny array. Such correlation can arise when the progeny come from a restricted number of pollen parents, as when a single pollen load is deposited by an insect. The determination of the maternal genotype by likelihood ratio from such an array may be biased if it assumed that outcrosses are half sibs rather than full sibs. This misclassification leads to estimates of outcrossing biased downwards. The correlation differs from that due to population structure in that the misclassification yields excessive numbers of heterozyous progeny in the apparently heterozygous maternal class. Schoen and Clegg (1984) developed a model for families with outcrosses from a single male parent for each progeny array, and the modified procedure for estimating the fraction of self pollinated seed in such cases. For tree species, it may be simpler to avoid if possible the complexities of shared paternity by collecting fruits from widely spaced branches, since these are likely to arise from separate pollinations.

5. Differential male fertility

Considerable theoretical interest has recently emerged in the extent to which plants differ in their contribution through male gametes, to the next generation (see Ross and Gregorius, 1983). Aside from the importance of such differentials to the evolution of breeding systems, is the practical significance of such variation. Thus in seed orchards, if only one clone were to provide all the successful pollen in seed production, the progeny generation would be subject to biased gene frequencies and elevated levels of inbreeding. (The progeny in this case would be a mixture of half- and full-sibs). A similar problem can arise in conservation of rare species such as Eucalyptus caesia, (Moran and Hopper, 1983), where natural populations are of very restricted size. Indeed the degree of inbreeding arising from such biases may be greater than that due to the self-fertilized component of predominantly outcrossed species.

A major issue concerns the extent to which such variation is related to genotype. Male fertility variance among individuals in natural populations might be anticipated; just as individual plants can show conspicuous variation in seed production. Presumably the genetic component for such variance is under intense selective pressure and could be rapidly exhausted unless opposed by countervailing pressures, as considered by sex allocation theory (Charnov, 1982). In natural populations, such pressures may be finely adjusted. In contrast, plantations and seed orchards may consist of individuals of disparate origin and phenology for which fertility variance may be extreme.

In relating variation to genotype three experimental situations present themselves. In populations with a finite number of distinct known genotypes, it is possible to disentangle the paternity of an array of seeds, given a sufficiently large battery of polymorphic marker loci (Ellstrand, 1984). The increased power of combined megagametophyte-seed scores in gymnosperms makes them ideal for this purpose. Genetic determination of paternity proceeds by exclusion, locus-by-locus, ideally so that all but one possible genotype is left as the potential source of pollen. Even then isolation has to be assumed. However, ambiguous paternity may be a frequent problem. Ritland (1983) formulated a likelihood approach for an estimate for any individual progeny.

The next problem is how to compute the male contributions summed over all progeny. It may be simplest to attribute ambiguous progeny fractionally to all possible male parents in proportion to their likelihoods. Such a routine would not take account of an observation that a genotype achieves a nonzero estimated contribution without ever being proved as a male parent by being involved in a unique attribution.

A second procedure, pioneered by Horovitz and Harding (1972), is to use morphological polymorphisms (such as flower colour or heterostyly) in natural or contrived populations, to test for differential male fertility, among the different morphs (Barrett et al, 1983).

Finally one may hope that differential male fertility might be reflected in hermaphrodites as a shift in allele frequency in the pollen as compared with parental population (Allard et al, 1977; Clegg, 1983). It should be noted however that relatively intense selection can occur at the diploid level without a noticable shift in allele frequencies.

CURRENT TOPICS FOR RESEARCH

The problem of measuring mating systems with genetic markers in progeny arrays can be viewed as how to transform the frequencies of observable outcrosses (X,Y) into meaningful quantities or processes - conventionally into outcrossing rates and gamete allele frequencies. We can call these the t-effects and the p-effects. Given that a number of studies have been published, one may well ask what are the more worthwhile questions upon which future work should focus.

The t-effects

Notwithstanding the published estimates of t in forest trees, it is still a real issue as to whether outbreeders show appreciable fractions of self-pollinated progeny in nature. Alternatively, is it valid to assume that selfing is essentially zero, and any nonzero estimate arises from bias (such as Wahlund effects)? Further, what circumstances lead to higher levels of selfing?

Moran and Bell (1983) summarized the several outcrossing estimates for Eucalyptus species (Table 5). In sum, these estimates are remarkably consistent in showing that partial selfing occurs in these animal-pollinated species. The outcrossing estimates are consistently less than those found for wind-pollinated conifers.

If appreciable selfing is found to occur, the important question becomes, can we pinpoint the stages of the life-cycle at which selection against inbreds is evident (Harding, 1975). Studies of seed harvest effects (such as Table 3), of embryo competition in gymnosperms, and of genotype changes during the life cycle could uncover the effect of selection. Stern and Roche (1974) have set out a detailed scheme for the gymnosperm life cycle which is a useful focus.

Table 5. Mean estimates of the outcrossing rate ($\hat{t}$) in 10 Eucalyptus species (from Moran and Bell, 1983)

Species	No. of populations	No. of loci	t	Source
E. obliqua	4	3	0.76	Brown et al. (1975)
E. pauciflora	3	7	0.70	Phillips & Brown (1977)
E. delegatensis	4	3	0.79	Moran and Brown (1980)
E. regnans	1	4	0.69	Moran (unpubl.)
E. stellulata	1	3	0.77	Moran (unpubl.)
E. stoatei	1	3	0.82	Hopper & Moran (1981)
E. kitsoniana	2	3	0.77	Fripp (1982)
E. citriodora	1	3	0.86	Yeh et al. (1983)
E. grandis	2	6	0.84	Bell (unpubl.)
E. saligna	1	6	0.77	Bell (unpubl.)

Differences among individuals in outcrossing rate are an important component of variation which has received little attention. Such variation should not be confused with variance in female fecundity (or seed production).

The p-effects

One important source of p-effects is that of subpopulation structure. This may masquerade in estimates as increased levels of apparent selfing, yet it is useful to hold to a basic difference in conception between real selfing (including geitonogamy, or even crosses between plants of the same clone) versus 'effective' selfing. In general p-effects are concerned with the source of the effective non-self pollen - to what extent is it contaminant, restricted, or biased. A particularly interesting focus of research is the question to what extent is there variation among individual plants and among genotypes in populations of forest trees, in their male fertility or contribution to the outcrossing pollen? At seed harvest, the female contributions to a seed bulk can be readily controlled - in marked contrast to the lack of control over male fertility variance. Further, in seed orchards such variance could be more noticable than in natural populations because planted populations may consist of selected clones of disparate origin, unlikely to be in equilibrium. As already discussed, the careful analysis of p-effects requires the construction of more complex, less general models, incorporating more features of the reproductive biology of the species.

The t-effects as opposed to the p-effects

Further we need to know the relative importance of variation in t-effects and p-effects, and their degree of interrelationship. What is the variance in selfing, compared with the variance in male fertility? Do the individuals or genotypes with higher rates of selfing have lower male fertility?

Significance of mating system studies

We are seeing a growth in studies of plant mating systems, particularly in forest trees. Studies based on markers may be descriptive (aimed at detecing what has happened in one mating cycle), or manipulative (testing what can happen in contrived populations). The overall aim is to understand the dynamic processes which result in the progeny generation, and to specify this in parameters attributable to the parental generations. The practical significance of such studies cannot be overemphasized in forest genetics - where the long term cost of producing poor seedling generations is substantial. The information is fundamental to the management of seed orchards so that the hard-won gains of selection will not be lost on inferior seed quality.

There are also important implications for the management of natural stands. Selective logging has been mainly debated in terms of dysgenic selection. Yet the effects on variation in mating system induced by selective logging may also be significant. Thus when outcrossing rates are density dependent, isolated individuals left after logging may be highly self-fertilized.

Third, an appreciation of the mating system is needed for defining optimal conservation strategies. We need to know in more quantitative terms, what are effective population sizes, neighborhood sizes, and migration rates, to answer questions about strategies (Brown and Moran, 1980). Only when these parameters are known with greater precision can a truly genetic management of biological resources be undertaken.

REFERENCES

Allard, R.W., Kahler, A.L., and Clegg, M.T. 1977. Estimation of mating cycle components of selection in plants. In: Measuring Selection in Natural Populations. (ed. Christiansen, F.B. and Fenchel, T.M.) pp. 1-19, Springer-Verlag, Berlin.

Bailey, N.T.J. 1951. Testing the solubility of maximum likelihood equations in the routine application of scoring methods.
Biometrics 7: 268-274.

Barrett, S.C.H., Price, S.D. and Shore, J.S. 1983. Male fertility and anisoplethic population structure in tristylous Pontederia cordata. Evolution 37:745-759.

Brown, A.H.D., Matheson, A.C., and Eldridge, K.G. 1975. Estimation of the mating system of Eucalyptus obliqua L'Herit. using allozyme polymorphisms.
Aust. J. Bot. 23: 931-949.

Brown, A.H.D., and Moran, G.F. 1980. Isozymes and the genetic resources of forest trees. In: Isozymes of North American Forest Trees and Forest Insects. (ed. Conkle, M.T.) pp. 1-10, U.S.D.A.

Charnov, E.L. 1982. The Theory of Sex Allocation. Princeton Univ. Press., N.J.

Cheliak, W.M., Morgan, K., Strobeck, C., Yeh, F.C.H., and Danick, B.P. 1983. Estimation of mating system parameters in plant populations using the EM algorithm.
Theor. Appl. Genet. 65: 157-161.

Clegg, M.T. (1983). Detection and measurement of natural selection. In: Isozymes in Plant Genetics and Breeding, Part A. (ed. Tanksley, S.D. and Orton, T.J.) pp. 241-255. Elsevier, Amsterdam.

Ellstrand, N.C. 1984. Multiple paternity within the fruits of the wild radish Raphanus sativus.
Am. Nat. 123: 819-828.

Ellstrand, N.C., and Foster, K.W. 1983. Impact of population structure on the apparent outcrossing rate of grain sorghum (Sorghum bicolor).
Theor. Appl. Genet. 66: 323-327.

Ennos, R.A., and Clegg, M.T. 1982. Effect of population substructuring on estimates of outcrossing rate in plant populations.
Heredity 48: 283-292.

Fripp, Y.J. 1982. Allozyme variation and mating system in two populations of Eucalyptus kitsoniana (Leuhm) Maiden.
Aust. For. Res. 13: 1-10.

Fyfe, J.L., and Bailey, N.T.J. 1951. Plant breeding studies in leguminous forage crops. 1. Natural cross-breeding in winter beans.
J. Agric. Sci. 41: 371-378.

Green, A.G., Brown, A.H.D., and Oram, R.N. 1980. Determination of outcrossing in a breeding population of Lupinus albus L.
Z. Pflanzenzucht. 84: 181-191.

Hamrick, J.L. 1982. Plant population genetics and evolution.
Amer. J. Bot. 69: 1685-1693.

Harding, J. and Tucker, C.L. 1964. Quantitative studies on mating systems. 1. Evidence for the non-randomness of outcrossing in Phaseolus lunatus.
Heredity 19: 369-381.

Harding, J. 1975. Models for gametic competition and self fertilization as components of natural selection in populations of higher plants. In: Gametic Competition in Plants and Animals. (ed. Mulcahy, D.L.) pp. 243-255. North Holland Publ. Co., Amsterdam.

Hopper, S.D., and Moran, G.F. 1981. Bird pollination and the mating system of Eucalyptus stoatei.
Aust. J. Bot. 29: 625-638.

Horovitz, A., and Harding, J. 1972. The concept of male outcrossing in hermaphrodite higher plants.
Heredity 29: 223-236.

Jones, D.F. 1916. Natural cross-pollination in the tomatoe. Science 43: 509-510.

Marshall, D.R., and Brown, A.H.D. 1974. Estimation of the level of apomixis in plant populations. Heredity 32: 321-333.

Moran, G.F., and Bell, J.C. 1983 Eucalyptus. In: Isozymes in Plant Genetics and Breeding, Part B (ed. Tanksley, S.D. and Orton, T.J.) pp. 423-441 Elsevier, Amsterdam.

Moran, G.F. and Brown, A.H.D. 1980. Temporal heterogeneity of outcrossing rates in alpine ash (Eucalyptus delegatensis R.T. Bak.). Theor. Appl. Genet. 57: 101-105.

Moran, G.F. and Hopper, S.D. 1983. Genetic diversity and the insular population structure of the rare granite rock species, Eucalyptus caesia Benth. Aust. J. Bot. 31:161-172.

Phillips, M.A., and Brown, A.H.D. 1977. Mating system and hybridity in Eucalyptus pauciflora. Aust. J. Biol. Sci. 30: 337-344.

Ritland, K. 1983 Estimation of mating systems. In: Isozymes in Plant Genetics and Breeding, Part A. (ed. Tanksley, S.D. and Orton, T.J.) pp. 289-302. Elsevier, Amsterdam.

Ritland, K. 1984. The effective proportion of self-fertilization with consanguineous matings in inbred populations. Genetics 106: 139-152.

Ritland, K., and Jain, S.K. 1981. A model for the estimation of outcrossing rate and gene frequencies using n independent loci. Heredity 47: 35-52.

Ross, M.D. and Gregorius, H.-R. 1983. Outcrossing and sex function in hermaphrodites: a resource-allocation model. Am. Nat. 121: 204-222.

Schoen, D.J. and Clegg, M.T. 1984. Estimation of mating system parameters when outcrossing events are correlated. Proc. Natl. Acad. Sci. U.S. 81:5258-5262.

Shaw, D.V., and Allard, R.W. 1982. Estimation of outcrossing rates in Douglas-fir using isozyme markers. Theor. Appl. Genet. 62: 113-120.

Shaw, D.V., Kahler, A.L., and Allard, R.W. 1981. A multilocus estimator of mating system parameters in plant populations. Proc. Natl. Acad. Sci. U.S.A. 78: 1298-1302.

Stern, K., and Roche, L. 1974. Genetics of Forest Ecosystems. Springer-Verlag, Berlin.

Yeh, F.C., Brune, A., Cheliak, W.M., and Chipman, D.C. 1983. Mating system of Eucalyptus citriodora in a seed production area. Can. J. For. Res. 13:1051-1055.

UNDERSTANDING THE GENETIC STRUCTURE OF PLANT POPULATIONS: SOME OLD PROBLEMS AND A NEW APPROACH

J. L. Hamrick and Andrew Schnabel

ABSTRACT

The genetic structure of natural plant populations results from the interaction of selection, gene flow and genetic drift. Reviews of the plant allozyme literature demonstrate that the distribution of allozyme variation within and among plant populations is closely associated with the species' mating system, pollination ecology and seed dispersal mechanism. Yet, there are relatively few species for which dependable estimates of the mating system or of pollen or seed dispersal are available.

Estimates of plant mating systems based on the mixed mating model have provided insights into the breeding structure of a few species. There are, however, relatively few data available on temporal and spatial variation in the mating system. Also, the assumptions of the mixed mating model are often violated in natural populations. Finally, the mixed mating model is limited in its ability to provide information about the breeding structure of populations.

Our understanding of gene flow via pollen or seed in natural populations is poor at best. Estimation procedures based on pollinator movements have underestimated pollen movement by not adequately dealing with pollen carryover and do not measure the effective movement of genes. Procedures using genetic markers, although giving a more accurate measure of pollen flow in natural or artificial populations typically produce results which are limited in scope and generalizations are difficult.

The use of paternity analysis to identify the father of individual seeds or seedlings removes many of the problems inherent to estimates of gene flow or the mating system. Although requiring considerable effort, paternity analysis can determine several genetic parameters that have previously been difficult or impossible to measure. From such analyses a detailed picture of the mechanisms which interact to produce the genetic structure of plant populations can be obtained.

INTRODUCTION

Natural plant populations often consist of patches of individuals in which one genotype or allele predominates (Hamrick and Holden, 1979; Linhart et al. 1981; Schaal, 1975; Turkington and Harper, 1979). Although this spatial genetic structure may be the result of natural selection, Wright (1931, 1938, 1951) and others (e.g., Turner, et al, 1982) have shown that genetic subdivision can be the result of limited gene flow and low effective population size (N_e). Effective population size is influenced by several factors including population density, the species' mating system (proportion of selfing and outcrossing), and the variances of pollen and seed dispersal. Thus, for the most part, spatial genetic structure is a function of the ability of the species to disperse its genes via pollen and seed. The distribution of allozyme variation within and among plant populations (Table 1) generally supports theoretical predictions of the importance of pollen and seed dispersal (Loveless and Hamrick, 1984). Genetic structure at the population subdivision level is also consistent with these predictions (Brown, 1979, Loveless and Hamrick, 1984).

Given the importance of gene flow and the mating system in determining the genetic composition of populations, it is surprising that quantitative measures of these traits are largely wanting for most plant species. Furthermore, those estimates that are available cannot be easily generalized to other species or populations. Thus, for many plant species, including several of those of commercial value, there is little information on the parameters that interact to produce spatial genetic structure. Even less is known about temporal or spatial variation in these parameters (Hamrick, 1982).

In the following sections we examine the procedures that are commonly used to provide quantitative estimates of plant mating systems and gene flow. Special attention is placed on the assumptions that underlie these estimation procedures and examples are given indicating that often these assumptions are not met in natural populations. Also, many of these procedures cannot provide the information needed to interpret the genetic structure of plant populations. Finally, we suggest that an analytical technique, paternity analysis, has the potential to provide quantitative measures for several of the parameters that interact to produce genetic structure in plant populations.

ESTIMATES OF PLANT MATING SYSTEMS

The most commonly used procedure for estimating proportions of selfing and outcrossing in plant populations is the mixed mating model (Allard and Workman, 1963; Brown and Allard, 1970). In this model, the proportion of outcrossed

Table 1. The influence of plant breeding systems and seed dispersal mechanisms on levels of genetic diversity among populations (From Loveless and Hamrick, 1984).

	Number of Studies	Mean Diversity Among Populations (G_{ST})
Autogamous	39	.523
Annual	31	.560
Perennial	8	.329
Mixed Mating	48	.243
Outcrossed	76	.118
Animal	32	.187
Wind	44	.068
Gravity	59	.446
Animal-Attached	18	.398
Animal-Ingested	14	.332
Explosive	24	.262
Winged/Plumose	48	.079

progeny (t) arising from homozygous maternal plants is estimated by dividing the proportion (h) of heterozygous progeny (i.e., outcrossed) by the frequency of heterozygotes expected with complete outcrossing (t = 1). This latter parameter is equal to the frequency of the alternative allele (p). Thus,

$$t = \frac{h}{p}.$$

Clegg (1980) lists several basic assumptions of the mixed mating model. The first assumption is that all mating events are due either to random outcrossing (with probability t) or to self-fertilization (with probability s (= 1 - t)). The possibility of assortative matings or matings with nearby related individuals is not recognized by the model. The second assumption follows directly from the first and states that pollen allele frequencies are identical over all maternal plants. In other words, regardless of an individual's location, the same probabilities of fertilization by the various pollen types apply. This assumption is not met if there is a non-random distribution of genotypes within the population and if mating is more likely between nearest neighbors. Third, the rate of outcrossing is assumed to be independent of the maternal genotype. This assumption does not hold if one maternal genotype is more or less accepting of self-fertilization. Last, the model assumes that no selection affecting the marker loci intervenes between mating and the determination of progeny genotypes. It will be violated if embryos resulting from selfing or some other form of inbreeding have lower survivorship before or during germination than progeny derived from outcrossing.

Each of these assumptions is probably rarely satisfied by natural plant populations. Unlike the assumptions of some models, the assumptions of the mixed mating model are relatively easy to test. Furthermore, in the process of testing these assumptions significant insights may be gained into the biology of the species, in particular its genetic structure.

The first two assumptions are the most easily violated. Most populations of plants have a genetic structure; alleles and genotypes are spatially aggregated. If matings occur primarily between nearest neighbors, estimates of outcrossing for individual maternal plants will vary significantly and overall estimates of outcrossing for the population will be reduced (Wahlund effect). This phenomenon has been demonstrated to occur in the animal pollinated *Eucalyptus pauciflora* (Phillips and Brown, 1977) and for two species of bee pollinated *Ipomoea* (Ennos, 1981). In fact, Schoen and Clegg (1984) demonstrated that a mating model in which progeny result either from selfing or from fertilization by a single outcrossed father (i.e., full sibs) fits the *Ipomoea* data better than the traditional mixed mating model.

In a recent study (Hamrick and Smith, in prep.) of a lodgepole pine (*Pinus contorta*) population located at 3000 m on the Front Range of the Colorado Rocky Mountains, outcrossing rates were estimated for a sample of 130 trees. The four-locus genotype of each mature individual was determined by analyzing its needle tissue. Several seedlings from each tree were analyzed for these same four loci. To test whether the first two assumptions of the mixed mating model were valid for this population, a chi-squared analysis (Phillips and Brown, 1977) was performed on the progeny arrays of trees homozygous for each locus (Table 2). The results indicated that for two loci, ADH and SKDH, the proportion of heterozygous progeny varied significantly among the homozygous maternal trees indicating that there was a non-random distribution of allele or genotype frequencies in the population. The other two loci, FE and GOT1, did not have significant differences among the progeny arrays of individual plants. As predicted, estimates of outcrossing (Table 2) for ADH and SKDH were lower than those for FE and GOT1.

A second study (Hamrick and Smyth, in prep.) further illustrates how the non-random distribution of pollen among individuals affects our perception of the mating system. This study was carried out in an artificial population of musk thistle (*Carduus nutans*), an insect pollinated composite. The population consisted of equal numbers of individuals homozygous for the fast (FF) and slow (SS)

Table 2. Analysis of heterogeneity in the proportion of heterozygous progeny for a *Pinus contorta* population from the Colorado Front Range. Also given are estimates of outcrossing (t) for each of the four allozyme loci analysed (Data from Hamrick and Smith, in preparation)

Locus	Maternal Genotype	D.F.	X^2	Significance Level	t
FE	22	47	50.10	$P < .25$	1.001
	33	5	6.41	$P < .50$	
ADH	11	29	52.81	$P < .01$	.901
	44	7	15.99	$P < .05$	
SKDH	22	11	16.50	$P < .25$	.923
	33	20	40.10	$P < .01$	
GOT-1	22	65	77.64	$P < .10$	.995
	33	1	5.24	$P < .05$	

migrating alleles at a phosphoglucoisomerase locus (PGI). The population was designed so that the eight nearest neighbors of SS individuals varied in genetic composition from eight SS individuals to eight FF individuals. A similar configuration existed for FF individuals. When the progeny arrays of each maternal plant were analyzed (Table 3), the proportion of heterozygous progeny varied between the two homozygous genotypes ($\chi_1^2 = 107.52$; $P < .01$) and among individuals with the same genotype. The first result could be due to differential pollen production by the two genotypes or by different rates of outcrossing for the two genotypes (assumption 3). The observation of significant heterogeneity among individual plants of the same genotype indicates that individual plants perceive the pollen pool in very different ways (assumption 2).

Since some *C. nutans* individuals had several heads flowering simultaneously, it was also possible to determine whether pollen deposited on flowering heads belonging to the same individual was genetically uniform. The results (Table 4) indicate that the frequency of heterozygous progeny was quite uniform among the heads of some plants (8 and 26), whereas for others (17 and 48) the frequency of heterozygotes within each head was quite variable. It is likely that such fine scale heterogeneity results from the foraging pattern of individual bees, and that, in at least some cases, the head is pollinated by one or a very few visitors. Unfortunately, the mixed mating model is not capable of determining how many paternal individuals contributed to the fertilization of single heads or individuals.

Table 3. Progeny arrays of PGI(FF) and PGI(SS) maternal plants from an artificial population of musk thistle, *Carduus nutans*, (Hamrick and Smyth, in preparation).

	FF Maternal Plants				SS Maternal Plants		
	Progeny Genotype				Progeny Genotype		
Plant #	FF	FS	% Heterozygotes	Plant #	SS	FS	% Heterozygotes
8	85	17	16.7	14	42	38	47.5
17	50	36	41.9	22	53	26	32.9
36	62	26	29.5	26	47	83	63.8
37	95	37	28.0	32	45	38	45.8
51	52	14	21.2	41	33	54	62.1
78	32	6	15.8	48	33	74	69.2
83	40	3	7.0	56	27	36	57.1
				69	53	31	36.9
TOTALS	416	139	25.0	75	39	47	54.7
				90	28	12	30.0
				99	17	27	61.4
				TOTALS	417	466	52.8
	$X^2_6 = 28.07$				$X^2_{10} = 54.68$		
	$P < .01$				$P < .01$		

Table 4. Progeny arrays from individual heads of five selected *C. nutans* plants growing in a common garden (Hamrick and Smyth, in preparation).

Plant Number	Genotype	Flower Number	Progeny genotype Homo	Hetero	Percent Heterozygotes	Mean	X^2
8	FF	1	18	2	10.0		
		2	16	4	20.0		
		3	18	4	18.2	16.7	2.48
		4	18	2	10.0		N.S.
		5	15	5	25.0		
17	FF	1	17	3	15.9		
		2	4	18	81.8	30.0	22.29
		3	13	9	40.9		$P < .01$
		4	16	6	27.3		
26	SS	1	8	12	60.0		
		2	11	11	50.0		
		3	3	19	86.4	63.8	7.79
		4	10	12	54.5		N.S.
		5	8	14	63.6		
		6	7	15	68.2		
37	FF	1	19	3	13.6		
		2	11	11	50.0		
		3	16	6	27.3	28.0	9.65
		4	17	5	22.7		
		5	14	8	36.4		$P < .10$
		6	18	4	18.2		
48	SS	1	15	7	31.8		
		2	5	17	77.3	69.2	18.92
		3	5	16	76.2		
		4	3	19	86.4		$P < .01$
		5	5	15	75.0		

The final assumption concerns whether selection occurs between the time of fertilization and genetic analysis. Although the action of selection at this stage of the life cycle has received little attention, the existing evidence indicates that considerable selection can occur (Clegg et al., 1978; Sorensen, 1982). For instance, in species that are primarily outcrossed seed set is considerably lower when high levels of selfing occur (Hagman and Mikkola, 1963; Lindgren, 1975; Sorensen and Miles, 1974). The selfed seed presumably have lower rates of survival than do outcrossed seed (Sorensen, 1982). Smith and Hamrick (in prep.) demonstrated that in lodgepole pine selfed strobili produce 80% fewer filled seed than outcrossed strobili. By using allozyme markers and crosses of 50/50 mixtures of selfed and outcrossed pollen they also demonstrated that the progeny were approximately 50% more heterozygous than expectations based on random fertilization (Table 5). These studies confirm that outcrossed pollen grains have a higher probability of producing viable seedlings.

Selection against lodgepole progeny resulting from self-fertilization can also be inferred from the survival of dormant seed stored in serotinous cones (Hamrick, in prep.). Because the cones, which can be aged, remain closed on the trees until fire causes them to open, estimates of the levels of outcrossing could be obtained for an eleven-year period (1966-1976). Although there was considerable year to year variation, the overall trend was towards an increase in estimates of outcrossing for seed from the older cones (Figure 1). These results are best explained by assuming that seeds resulting from selfing have lower survival rates when stored for long periods in a dormant state. Similar patterns of mortality have been observed by Moran and Brown (1980) over a three year period for *Eucalyptus delegatensis* and by Cheliak (1985) over a four year period for *Pinus banksiana* .

We see, therefore, that the assumptions of the mixed mating model are commonly violated in natural plant populations. This may be especially serious in animal pollinated species and may lead to inaccurate estimates of outcrossing rates (Schoen and Clegg, 1984). A more serious problem is that the mixed mating model was not designed to describe the breeding structure of populations. To understand those factors that interact to produce genetic structure in plant populations it is necessary to determine the pattern of matings among individuals and to know the genetic relatedness of progeny produced by each individual.

Table 5. Results of controlled crosses which used a 1:1 mixture of selfed to outcrossed *Pinus contorta* pollen (Smith and Hamrick, in preparation)

	ADH	FE	GOT1	GOT2
Number of crosses	4	4	1	1
Expected Heterozygotes	96	251	12	12
Observed Heterozygotes	134	386	25	25
Ratio	1.40	1.54	2.08	2.08

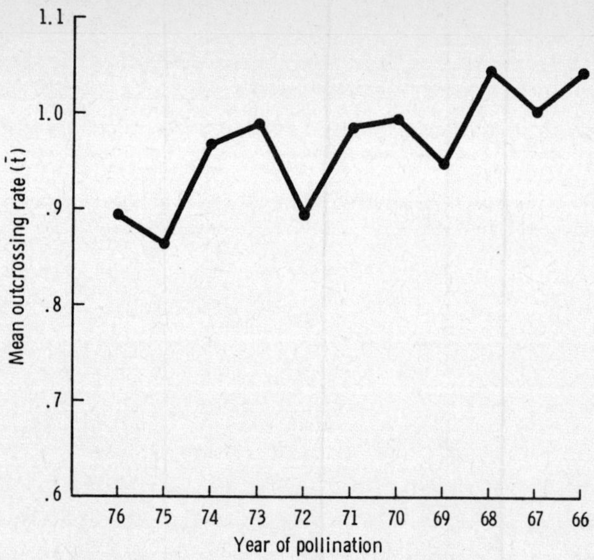

Figure 1. The mean outcrossing rate ($\bar{t}$) of a *Pinus contorta* population as a function of the year of pollination (from Hamrick and Smith, in prep).

ESTIMATES OF GENE FLOW

There are several methods that have been developed to estimate gene flow. The most common approach is to measure the movement of dispersal units (pollen, seed) or their dispersal vectors. In the latter case it is often assumed that the majority of the pollen picked up on the first plant is deposited after only one or two subsequent visits (Levin and Kerster, 1968; 1974). Since pollinator visits are typically between near neighbors, estimates of pollen movement have, in general, been low and have produced small values of N_e. More recent studies of pollen carryover (Thomson and Plowright, 1980; Waser and Price, 1982; Lertzman and Gass, 1983) contradict these assumptions by indicating that pollen can be deposited on flowers removed by several visits from the pollen source. Furthermore, the amount of pollen deposited varies greatly from visit to visit. Both pollen carryover and variable pollen deposition should have marked effects on genetic structure, but neither has been studied with this question in mind. Despite these problems, the movement of pollinators to and from several plants can be easily monitored. Thus, the pollination ecology of individuals that are located in a variety of densities, patch sizes or positions within patches (edge vs. center) can be examined. This is a distinct advantage of this technique over many of the approaches described below.

Following the actual movement of pollen or seed rather than their vectors is more difficult, but may be accomplished by marking pollen or seed with dyes or radioactive tracers or by measuring levels of pollen or seed deposition around isolated individuals. An example of this latter approach is that of Silen (1962), who examined pollen deposition patterns around a small cluster of Douglas fir (*Pseudotsuga menziesii*) in Oregon. The pattern of pollen deposition was strongly leptokurtic and levels of pollen deposition above background levels were found beyond 650 m from the source plants. Whereas this approach gives a more accurate description of pollen (or seed) movement than do studies based on the movements of vectors, it provides no information on fertilization success. Nor do we know what proportion of the pollen comes from outside the populations' boundaries. Conclusions based on studies of vector movements or pollen and seed dispersal have generally stated that gene movement within and among plant populations is low and of little evolutionary consequence (Ehrlich and Raven, 1969; Levin and Kerster, 1974).

Many of the problems discussed above do not arise if the movement of unique alleles is followed in natural or artificial populations. For instance, in a variety of crop plants, Bateman (1947) found low frequencies of single gene morphological markers at distances up to 170 m from their source. Muller (1977) used a unique LAP allele to measure gene flow in *Pinus silvestris*. He observed that the marker allele occurred in progeny of trees up to 80 m from the source. Likewise, Schaal (1980) used an electrophoretic marker to study pollen movement in a population of *Lupinus texensis*. The mean distance of gene movement was 1.8 m, approximately twice the estimate provided by studies of pollinator movement.

Since this method is dependent on finding individuals with unique alleles, it is not applicable to all species or populations. Therefore, genetic markers have more often been used in artificial populations. Typically, a source area containing the unique allele is established, and these individuals are surrounded with plants homozygous for an alternative allele. In an artificial population of *Carduus nutans*, Smyth and Hamrick (in prep.) found that effective pollen movement followed a leptokurtic distribution with the genetic marker moving at least 18 m from its source (Figure 2). Interestingly, the marker allele was patchily distributed among individuals and heads (Figure 3). Some plants, especially those further from the source, had no progeny with the marker whereas a third of the progeny of other, equally distant, individuals were heterozygous. Similar variation in marker allele deposition was seen among individual heads of the same plant. The patchy distribution of the marker is consistent with the pattern of pollen deposition inferred from the study of the mating system of *C. nutans* (see above), and is almost certainly due to the foraging behavior of the pollinators coupled with pollen carryover. Handel (1982) found similar patterns of pollen deposition in an artificial

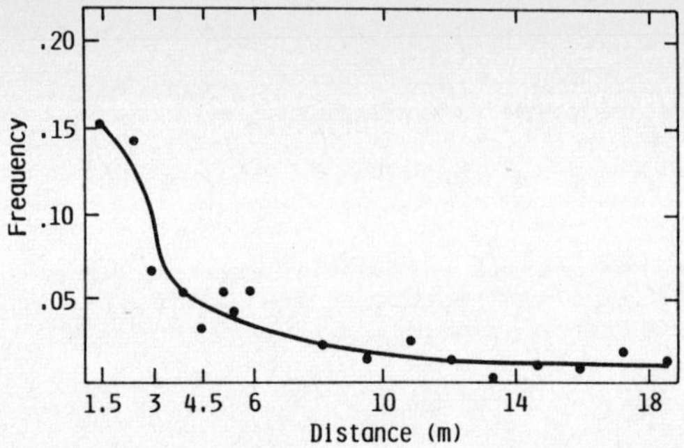

Figure 2. The mean frequency of a unique genetic marker (PGI^S allele) as a function of distance from the source plant in an artificial population of *Carduus nutans* (from Smyth and Hamrick in prep.).

population of *Cucumis melo*. Although not affecting the overall quantity of gene flow such a clumped distribution of alleles would have a marked effect on genetic structure.

In general, the use of genetic markers gives a more accurate picture of the effective movement of genes within and among populations. A serious problem, however, is the difficulty of finding individuals in natural populations that have a rare allele. At best only a few source plants will be available and as a result generalizations of the patterns of gene movement in species or populations may not be possible. The use of artificial populations has many of these same problems of replicability and the value of the results will be a function of the experimental design. Only rarely do studies of artificial populations consider variables such a dispersion, population shape or population density, factors that almost certainly are important influences on gene flow in natural populations.

Finally, Slatkin (1980) proposed a measure of gene flow based on allele frequency distributions within and among populations. This procedure provides a qualitative rather than a quantitative estimate of gene flow. It is valuable from a comparative standpoint for distinguishing among species with high, low or intermediate levels of gene flow. Its utility is illustrated by the work of Hamrick and Griswold (in prep.), who grouped species according to their breeding system and applied Slatkin's analysis to the allele frequency distributions in each species (Figure 4).

In conclusion, most of the currently used measures of gene flow in plant populations do not provide detailed or dependable estimates of gene movement. Conclusions, based largely on pollen or pollinator movement have been used to

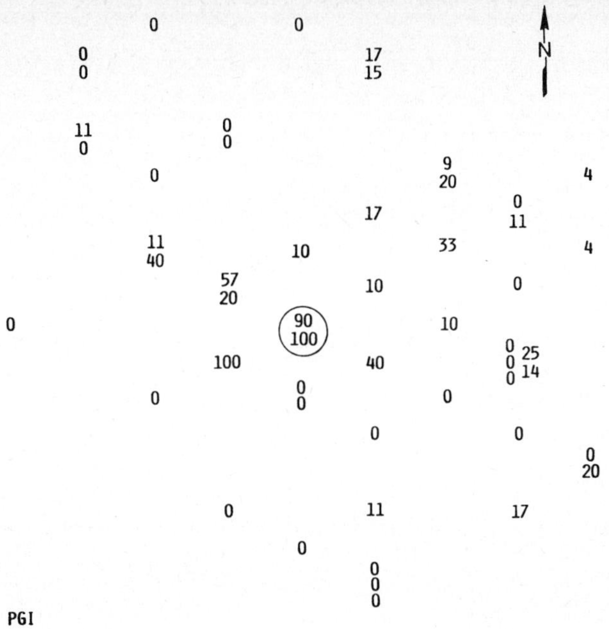

Figure 3. The actual distribution of a unique genetic marker (PGI^S allele) in an experimental population of *Carduus nutans*. The numbers represent the proportion of heterozygous progeny. Two or more numbers at a location are values for individual flowering heads. The source plant is circled. Note the marked heterogeneity in values among plants at the same distance and among heads within a plant (from Smyth and Hamrick, in prep.).

argue that gene movement in most plant species is limited (Ehrlich and Raven, 1969; Levin and Kerster, 1974) and that neighborhood sizes are small. The generality of these conclusions for a broad spectrum of plant species is called into question by the results in Table 1 and Figure 4. It is obvious that plant species vary greatly in their ability to transport genes. In order to obtain accurate, generally applicable estimates of gene flow we need to measure pollen and seed movement from several plants in a variety of populations. Furthermore, estimates of gene flow should be taken over several years so that temporal variation in gene movement can be quantified.

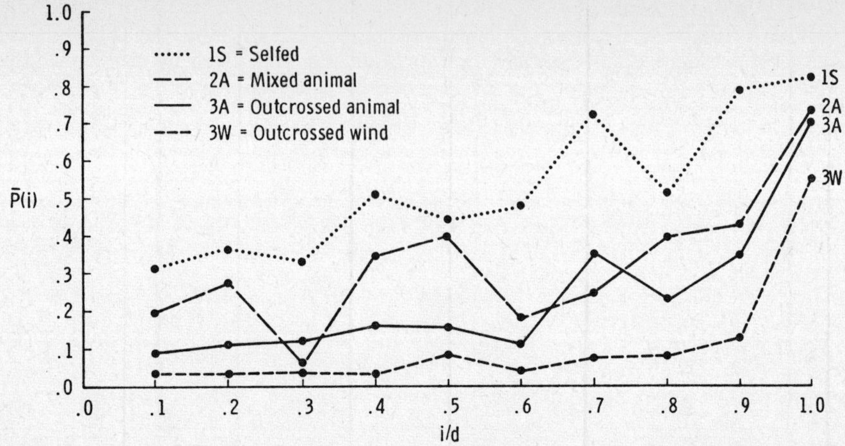

Figure 4. The relationship between the mean frequency of an allele in those populations in which it occurs (conditional average frequency, $\bar{p}_i$) and the relative occupany number (i/d; where i = the number of populations in which an allele is present and d = the total number of populations) for groups of plant species with different breeding systems (from Hamrick and Griswold, in prep.).

A NEW APPROACH - PATERNITY ANALYSIS

Many of the problems with our ability to adequately describe the breeding structure of natural plant populations or to measure gene movement within and among populations could be resolved if the paternity of individual progeny were known. The presence of large amounts of allozyme variation in populations of many plant species (Hamrick et al., 1979) provides a mechanism for the determination of paternity.

Two general approaches can be used. If the maternal and progeny genotypes are known for each polymorphic locus, a certain proportion of the potential fathers can be excluded with each locus. For example, suppose a A_1A_2 maternal plant has a A_2A_2 progeny. We know that both of the parental plants contributed a A_2 gamete. Potential paternal plants which do not have the A_2 allele are therefore excluded as the father of this individual (see Ellstrand, 1984 for a more complete explanation). As each locus is examined, a larger proportion of the potential fathers is excluded. The ability to exclude all but one paternal plant depends on the number of polymorphic loci, the number of alleles per locus, and the allele frequency distribution at each locus. Loci with several alleles in nearly equal frequencies provide the greatest resolution.

A second approach is a maximum likelihood analysis. Instead of excluding potential fathers, the individual that was most likely to have contributed the paternal gamete is identified. As with the exclusion analysis the discriminating power of this technique improves with increasing numbers of polymorphic loci and alleles and with equal allele frequencies. An assumption necessary for this approach is that the loci are not linked.

Uses of Paternity Analyses

Since all paternal plants within a population can be identified, the distance each successful pollen grain has moved can be determined. If several maternal plants are studied, a similar analysis can be applied to the dispersal of pollen away from paternal individuals. As a result, several parameters that were previously difficult to measure can be described in considerable detail. The proportion of effective pollen originating outside of the study area can be determined and exact distances of pollen movement within the study area can be measured. Furthermore, plant-to-plant variation for incoming (females) or outgoing (males) pollen dispersal can be estimated. By carefully selecting the location of the monitored plants or by manipulation of the population, the effects of density, dispersion and the location of individuals (edge vs center) on patterns of pollen movement can be described.

Actual rates of selfing and outcrossing are given by paternity analysis. If all fathers are identified, including the "maternal" plant, the actual number of progeny resulting from self-fertilization can be counted. Thus, it will be possible to separate inbreeding due to selfing from that due to consanguineous matings. It should also be possible to study the effects of population structure on selfing rates. Furthermore, if there are several seeds per flower or flowering head, it should be possible to determine whether selfing rates vary with the stage of flowering or flower position.

The genetic relatedness of progeny from a single maternal plant can be determined through paternity analysis. The mixed mating model is only capable of estimating the proportion of selfing, with the outcrossed progeny assumed to be the result of random mating (Clegg, 1980). Although this assumption is generally recognized not to be valid, estimates of the actual genetic relatedness of progeny have not been available. This information is especially important if seed and pollen dispersal are limited since considerable inbreeding could result if a population consists of patches of full-sibs.

The number of full-sib families (i.e., number of paternal individuals) within a single progeny array can be determined if all paternal plants can be identified. Also, the number of progeny belonging to each full-sib family can be counted. In species with multiple-seeded fruits, the number of fathers contributing pollen to each fruit can be determined (Ellstrand, 1984). A high percentage of progeny fertilized by a single paternal individual might indicate that a single pollinator visit produced the majority of the fertilizations.

If done in conjunction with studies of pollinator behavior paternity analyses can provide insights into phenomena such as pollen carryover. If the movement of a pollinator is followed and subsequent visits are prevented, the distribution of effective pollen deposition can be determined. Experiments can be designed to study the effects of the order of pollen deposition on fertilization. Pollen from known fathers could be placed on the stigmas at set intervals, and, by using paternity analyses, the number of fertilizations due to each father could be determined.

With paternity analyses the male reproductive success of each plant can be estimated by examining progeny from several plants. Accurate estimates of pollen production have been difficult to obtain and gave no indication of the plant's success in fertilization. By applying paternity analyses, the success of males in pollen competition studies can easily be examined. There has been considerable recent discussion of mate choice in plants (Willson and Burley, 1983). Documentation of mate choice in plants is difficult without the use of genetic markers. With single gene markers the success of several males could not be differentiated. By using paternity analyses, predetermined pollen mixtures of several randomly collected males can be applied to the stigmas of several females and the mating success of each male can be demonstrated.

Seed dispersal can also be studied by the same techniques used for paternity analyses. In this case the genotype of established seedlings are determined and potential parents are excluded. The actual maternal plant can not be distinguished for hermaphroditic or monoecious species, but this could be accomplished with dioecious species.

Paternity analyses could also be used to gain insight into patterns of matings in hybrid zones. In areas where hybridization and introgression are thought to occur paternity analyses could be used to determine if hybrid individuals are more likely to mate with other hybrids or whether they mate more often with individuals of either of the parental taxa.

Paternity Analysis of a Population of *Gleditsia triacanthos*

Gleditsia triacanthos (Fabaceae) is a dioecious tree that commonly colonizes old fields and pastures in Kansas. It is pollinated by insects, predominately bees (Schnabel, pers. obs.). In a population located seven miles northeast of Lawrence, Kansas, the spatial location and the ten-locus genotype of every mature male individual was determined. Within this area a single centrally located female plant was chosen for preliminary analysis. The ten-locus allozyme genotypes for 52 progeny from this individual were determined, and a paternity exclusion analysis was applied to the data. Of the 52 progeny, three were determined to have a father that was not in the study area. A single father was indicated for 17 progeny and the remaining 32 progeny could have been fathered by more than one of the 92 male trees. In most cases, however, more than 90% of the potential fathers were excluded.

The results indicate that eight paternal plants had sired the 17 progeny for which a single father could be identified (Figure 5). Two male plants sired four progeny and three males had two progeny each. The mean distance of pollen movement for these 17 progeny was 49.3 m.

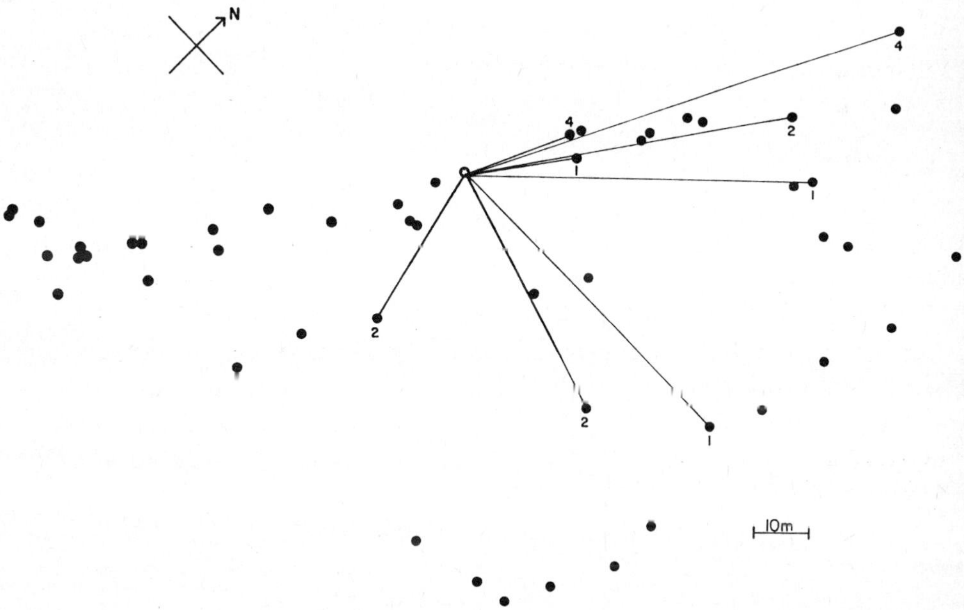

Figure 5. The movement of pollen from male *Gleditsia triacanthos* to a single female. The location of male individuals is indicated by the closed circles. The fathers of 17 of 52 progeny could be determined by paternity exclusion analysis. Numbers adjacent to the tree locations indicate the number of progeny sired by each male.

CONCLUDING REMARKS

Many plant species may not maintain levels of allozyme variation sufficient for the application of paternity analyses. The results presented above for *Gleditsia* indicate that ten polymorphic loci will probably be the minimum needed. Several of the loci in this population had two alleles and allele frequencies were strongly skewed. At least two additional highly polymorphic loci would be needed to resolve the paternity of several of the progeny in this population. The application of maximum likelihood techniques to these data might also appreciably improve the resolution. A second disadvantage is that to describe thoroughly the breeding structure within a population several (50+) progeny from several maternal plants should be sampled. If more than one population is studied, the sample sizes will become quite large.

We feel, however, that the advantages gained by paternity analysis far outweigh any disadvantages. The analysis provides more exact information on several population parameters for which only estimates were previously available (gene flow, mating systems). In addition, paternity analysis provides quantitative values for parameters we could only guess at previously (genetic relatedness of progeny, number of paternal plants fertilizing each maternal plant or fruit, male reproductive values, mate choice, and seed shadows). Finally, an important advantage is that several aspects of the breeding structure of populations can be described from a single electrophoretic analysis. It appears that for those species where paternity analyses can be successfully applied, we will finally begin to understand those mechanisms that interact to influence the genetic structure of populations.

ACKNOWLEDGMENTS

We wish to thank Dr. M.D. Loveless for her contributions to the ideas expressed in this paper. Without her intellectual input it is doubtful that this paper would exist. We also wish to thank Linda Vescio, Cathy Gorman and Donna Frank for their technical assistance on much of the research reported herein. Portions of the research were supported by EPA Grant R-805455030 and National Science Foundation Grant DEB 7922190.

REFERENCES

Allard, R. W. and P. L. Workman. 1963. Population studies in pedominantly self-pollinated species. IV. Seasonal fluctuations in estimated values of genetic parameters in lima bean populations. Evol. 17:470-480.

Bateman, A. J. 1947. Contamination in seed crops. III. Relation with isolation distance. Heredity 1:303-336.

Brown, A. H. D. 1979. Enzyme polymorphism in plant populations. Theor. Pop. Biol. 15:1-42.

Brown, A. H. D. and R. W. Allard. 1970. Estimation of the mating system in open-pollinated maize populations using isozyme polymorphisms. Genetics 66:133-145.

Cheliak, W. M. 1985. Temporal variation in the mating system of jack pine, *Pinus barksiana*. Genetics, in press.

Clegg, M. T. 1980. Measuring plant mating systems. BioScience 30:814-818.

Clegg, M. T., A. L. Kahler, and R. W. Allard. 1978. Estimation of life cycle components of selection in an experimental plant population. Genetics 89:765-792.

Ehrlich, P. R. and P. H. Raven. 1969. Differentiation of populations. Science 165:1228-1231.

Ellstrand, N. C. 1984. Multiple paternity within the fruits of the wild radish, *Raphanus sativus*. Am. Nat. 123:819-828.

Ennos, R. A. 1981. Quantitative studies of the mating system in two sympatric species of *Ipomoea* (Convolvulaceae). Genetica 57:93-98.

Hagman, M. and L. Mikkola. 1963. Observations on cross- self- and inter-specific pollinations in *Pinus peuce* Griseb. Silv. Genet. 12:73-79.

Hamrick, J. L. 1982. Plant population genetics and evolution. Am. J. Bot. 1685-1693.

Hamrick, J. L. and G. B. Griswold. In prep. Association between Slatkin's measure of gene flow and the dispersal ability of plant species.

Hamrick, J. L. and L. R. Holden. 1979. Influence of microhabitat heterogeneity on gene frequency distribution and gametic phase disequilibrium in *Avena barbata*. Evolution 33:521-533.

Hamrick, J. L., Y. B. Linhart, and J. B. Mitton. 1979. Relationships between life history characteristics and electrophoretically-detectable genetic variation in plants. Ann. Rev. Ecol. Syst. 10:173-200.

Hamrick, J. L. and C. C. Smith. In prep. Temporal variation in estimates of the mating system in lodgepole pine, *Pinus contorta*.

Hamrick, J. L. and C. A. Smyth. In prep. Inter- and intraplant variation for estimates of the mating system in an experimental population of *Carduus nutans*.

Handel, S. N. 1982. Dynamics of gene flow in an experimental population of *Cucumis melo* (Cucurbitaceae). Am. J. Bot. 69:1538-1546.

Lertzman, K. P. and C. L. Gass. 1983. Alternative models of pollen transfer. *In*: Handbook of Experimental Pollination Biology, C. E. Jones and R. J. Little (eds.). Van Nostrand Reinhold, N.Y. pp. 474-489.

Levin, D. A. and H. W. Kerster. 1968. Local gene disperal in *Phlox*. Evol. 22:130-139.

Levin, D. A. and H. W. Kerster. 1974. Gene flow in seed plants. Evol. Biol. 7:139-220.

Lindgren, D. 1975. The relationship between self-fertilization, empty seeds and seeds originating from selfing as a consequence of polyembryony. Studia Forestalia Suecica Nr. 126.

Linhart, Y. B., J. B. Mitton, K. B. Sturgeon and M. L. Davis. 1981. Genetic variation in space and time in a population of ponderosa pine. Heredity 46:407-426.

Loveless, M. D. and J. L. Hamrick. 1984. Ecological determinants of genetic structure in plant populations. Ann. Rev. Ecol. Syst. 15:65-95.

Moran, G. F. and A. H. D. Brown. 1980. Temporal heterogeneity of outcrossing rates in alpine ash (*Eucalyptus delegatensis* R. T. Bak.). Theor. Appl. Genet. 57:101-105.

Muller, G. 1977. Cross-fertilization in a conifer stand inferred from enzyme gene-markers in seeds. Silv. Genet. 26:223-226.

Phillips, M. A. and A. H. D. Brown. 1977. Mating system and hybridity in *Eucalyptus pauciflora*. Aust. J. Biol. Sci. 30:337-344.

Schaal, B. A. 1975. Population structure and local differentiation in *Liatris cylindraceae*. Am. Nat. 109:511-528.

Schaal, B. A. 1980. Measurement of gene flow in *Lupinus texensis*. Nature 284:450-451.

Schoen, D. J. and M. T. Clegg. 1984. Estimation of mating system parameters when outcrossing events are correlated. Proc. Nat. Acad. Sci. 81:5258-5262.

Silen, R. R. 1962. Pollen dispersal considerations for douglas-fir (*Pseudotsuga menziesii*). J. For. 60:790-795.

Slatkin, M. 1981. Estimating levels of gene flow in natural populations. Genetics 99:323-335.

Smith, C. C. and J. L. Hamrick. In prep. Experimental studies on the effects of selfing in lodgepole pine, *Pinus contorta*.

Smyth, C. A. and J. L. Hamrick. In prep. Realized gene flow via pollen in artificial populations of musk thistle, *Carduus nutans*.

Sorensen, F. C. 1982. The roles of polyembryonyonal embryo viability in the genetic system of conifers. Evol. 36:725-733.

Sorensen, F. C. and R. S. Miles. 1982. Inbreeding depression in height, height growth, and survival of douglas-fir, ponderosa pine, and nobile fir to 10 years of age. For. Sci. 28:283-292.

Thomson, J. D. and R. C. Plowright. 1980. Pollen carryover, vector rewards, and pollinator behavior with special reference to *Diervilla lonicera*. Oecologia 46:68-74.

Turkington, R. and J. L. Harper. 1979. The growth, distribution and neighbour relationships of *Trifolium repens* in a permanent pasture. IV. Fine-scale biotic differentiation. J. Ecol. 67:245-254.

Turner, M. E., J. C. Stephens and W. W. Anderson. 1982. Homozygosity and patch structure in plant populations as a result of nearest-neighbor pollination. Proc. Nat. Acad. Sci. 79:203-207.

Waser, N. M. and M. V. Price. 1982. A comparison of pollen and fluorescent dye carry-over by natural pollinators of *Ipomopsis aggregata* . Ecology 63:1168-1172.

Willson, M. F. and N. Burley. 1983. Mate choice in plants. Tactics, mechanisms and consequences. Princeton University Press, Princeton, N.J. 251 pp.

Wright, S. 1931. Evolution in Mendelian populations. Genetics 16:97-159.

Wright, S. 1938. Size of population and breeding structure in relation to evolution. Science 87:430-431.

Wright, S. 1951. The genetical structure of populations. Ann. Eugenics 15:323-354.

INBREEDING AND SELECTION IN NATURAL POPULATIONS

Philip W. Hedrick

Abstract

Partial inbreeding due to sibmating can reduce heterozygosity beyond that resulting from partial self-fertilization alone. A negative autocorrelation in inbreeding over time, results in higher heterozygosity than when there is no autocorrelation. High selfing can result in a greater probability of polymorphism when there are selective differences over space. When two loci are considered simultaneously, inbreeding may increase or decrease the likelihood of polymorphism. Genetic hitchhiking may be basis of genetic change in highly selfed plants.

Introduction

A substantial amount of effort and thought is being focused on obtaining accurate and unbiased estimates of the mating system and gene flow (e.g., Brown et al., 1985; Hamrick and Schnabel, 1985). An important extension of these measurements is to evaluate the significance of these levels of inbreeding and gene flow in a general evolutionary context. For example, if we combine inbreeding and gene flow with various modes of selection, what might we predict about patterns of genetic variation over space and time?

My purpose here is to focus on some models that will illustrate the effects of inbreeding and selection, and to some extent gene flow and genetic drift, on the extent and pattern of genetic variation. I will begin by examining a single-locus model with partial inbreeding. Then I will discuss in sequence selection and inbreeding, selection and gene flow, and finally all three factors jointly. In the second part, I will examine more than one locus, first a two-locus neutrality model and then two-locus selection models with partial inbreeding. Finally, I will present a three-locus genetic hitchhiking model with partial inbreeding, selection, and gene flow.

Single Locus

Effect of inbreeding: Inbreeding can result from self-fertilization or from mating between relatives (e.g., Ritland, 1984). Estimates of the proportion of self-fertilization vary greatly among different plant species. For example, in conifers, they are generally low, of the order of 0.10, while in some grasses they approach 1.0. Estimates of inbreeding from between relative matings are generally unknown although, because of low gene flow in many plants, such matings may be common. It is very possible that the proportion of matings between sibs, parents and offspring, or cousins may be as high or higher than the proportion of self-fertilization in many species.

What effect does inbreeding have on the genetic variation in a population? To determine this, let us examine a locus $\underline{A}$ with two alleles, $\underline{A}_1$ and $\underline{A}_2$, having frequencies of $\underline{p}$ and $\underline{q}$, respectively. First, if we assume that a population reproduces a proportion $\underline{S}$ of its zygotes by self-fertilization and the remainder $1 - \underline{S}$ by random mating, then the equilibrium heterozygosity in such a mixed-mating model is

$$H_e = 4pq\left(\frac{1 - S}{2 - S}\right). \qquad [1]$$

Second, if we assume that another population reproduces a proportion $\underline{S}^*$ by full-sib mating and $1 - \underline{S}^*$ by random mating, then the equilibrium heterozygosity is

$$H_e = 8pq\left(\frac{1 - S^*}{4 - 3S^*}\right) \qquad [2]$$

(Li, 1976). Last, if we assume that a population has both selfing and full-sib mating, with the constraint that $\underline{S} + \underline{S}^* \leqq 1$, then the equilibrium heterozygosity is

$$H_e = 8pq\left(\frac{1 - S - S^*}{4 - 2S - 3S^*}\right). \qquad [3]$$

I should note that these inbreeding equilibrium levels of heterozygosity are reached fairly quickly, i.e., within five to ten generations of constant partial inbreeding, they are very close to the equilibrium value.

Let us compare the equilibrium heterozygosity for partial selfing or partial full-sib mating relative to the Hardy-Weinberg heterozygosity, $2\underline{pq}$ (see Figure 1). Notice that the heterozygosity is not reduced very much relative to Hardy-Weinberg proportions at low levels of inbreeding. For example, with 0.25 selfing or full-sib mating, the heterozygosity is 0.857 and 0.923 of that in a random-mating population. At increased levels of inbreeding, both types of inbreeding greatly reduce heterozygosity.

If there is more than one type of inbreeding, then the cumulative effect is greater than the sum of the individual effects. For example, if a population had 0.25 selfing and 0.25 full-sib mating, the equilibrium heterozygosity is 0.727 that of a random-mating population while the sum of the individual effects reduces heterozygosity to 0.78 of Hardy-Weinberg proportions. This phenomenon is due to the convex nature of the curves in Figure 1. Additional types of inbreeding such as half-sib matings and parent-offspring matings further decreases heterozygosity in a similar manner (Hedrick, 1985a).

In most cases, estimates of the selfing proportion are more easily obtained by direct measurement than are estimates of other matings between relatives. Assuming that both selfing and full-sib mating are occuring in a population and that the proportion of selfing can be measured directly, what level of full-sib mating would be necessary to account for an observed heterozygosity below that predicted by selfing alone? Using expressions (1) and (3) and assuming $\underline{S}_I$ is the level of selfing necessary by itself to explain an observed heterozygosity, an indirect

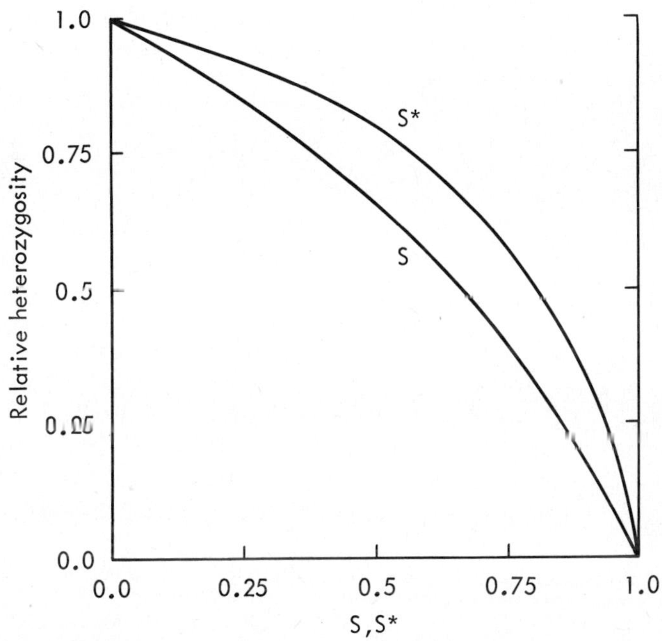

Figure 1. The relative heterozygosity for different levels of partial self fertilization ($\underline{S}$) and partial full-sib mating ($\underline{S}$*).

measure of selfing, then

$$S^* = \frac{2(S_I - S_D)}{1 + S_I} \qquad [4]$$

where $\underline{S}_D$ is the selfing level measured directly (Brown, 1979). For example, if $\underline{S}_D$ is estimated to be 0.1 and the observed heterozygosity is 0.889 that of

Hardy-Weinberg proportions, making $\underline{S}_I = 0.2$, then 0.167 full-sib mating in the population would explain this deficiency of heterozygosity. Of course, S* in this case represents the cumulative effects of all consanguineous matings except the proportion of selfing (see also Hedrick, 1985a).

Brown (1979) discusses what he terms "the heterozygosity paradox" in which species with high selfing often show an excess of heterozygosity from that expected using inbreeding equilibrium and species with low selfing have a deficiency of heterozygosity. Perhaps inbreeding between other relatives as discussed here is important in causing species with low selfing rates to have such a deficiency in heterozygosity.

Variable inbreeding over time. There is evidence that the extent of selfing and presumably other types of inbreeding varies over time (generations). For example in newly founded populations, the level of inbreeding, such as selfing, sib mating, or parent-offspring mating, may be high. After the colony is established, inbreeding may be much lower.

Whenever there is no inbreeding in one generation, heterozygosity returns to the Hardy-Weinberg proportion in the next generation. On the other hand, for complete inbreeding, heterozygosity is eliminated over several generations, and in general by five to ten generations, it is very low to zero. Nei (1975) and Brown (1979) have discussed this phenomenon and suggested because of the asymmetrical rate of return to inbreeding equilibrium that selfing estimates in a few generations may lead to the observed heterozygosity being higher than expected.

When the selfing proportion varies over time and there is zero autocorrelation between selfing levels, the equilibrium genotypic proportions, actually the expected heterozygosity over a long period of time, are obtained by using the mean selfing level (Brown, 1979; see also Allard et al., 1968). In other words, if $\underline{\bar{S}}$ is the mean level of selfing over generations in a population, then the equilibrium heterozygosity is

$$H_e = 4pq \frac{1 - \bar{S}}{2 - \bar{S}} . \qquad [5]$$

However, if there is a negative autocorrelation between inbreeding levels in different generations, the heterozygosity is somewhat higher than given by [5] while if there is a positive autocorrelation, it is lower (Hedrick, 1985b).

As an extreme case, let us assume that selfing levels vary between two values. When there is an autocorrelation of 1 between these levels, really infinitely long runs at each selfing level, then the equilibrium heterozygosity is

$$H_e = 2pq\left(\frac{1 - S_1}{2 - S_1} + \frac{1 - S_2}{2 - S_2}\right) \qquad [6]$$

where S_1 and S_2 are the two different levels of selfing. In other words, the overall equilibrium heterozygosity is the arithmetic average of the equilibrium heterozygosity for each level. On the other hand, when the autocorrelation is -1, a switch in selfing level every generation, then the heterozygosity is

$$H_e = 2pq(1 - \bar{S}) + \tfrac{1}{4}(S_1 H_2 + S_1 H_2) \ . \qquad [7a]$$

If we assume $S_1 = 0$, then $H_1 = 2pq$ and [7a] becomes

$$H_e = 2pq(1 - \tfrac{1}{4}S_2) \ . \qquad [7b]$$

Obviously, the overall heterozygosity is closer in this case to that for the lower level of selfing.

Table 1 gives three numerical examples to illustrate these effects. The most extreme example possible is given in the first two columns in which selfing varies between 0.0 and 1.0. In this case $\bar{S} = 0.5$ and the equilibrium heterozygosity for a zero autocorrelation is 0.667 that of Hardy-Weinberg proportions. However, when the autocorrelation is -1, the heterozygosity is elevated to 0.75 of Hardy Weinberg and when it is 1, it is lowered to 0.5. Two other examples are given for a predominant outcrosser, such as most conifers, and an predominant inbreeder, such as some grasses. Notice that the autocorrelation has only a small effect on heterozygosity in these cases. It is possible that a negative autocorrelation between selfing rates due perhaps to weather may explain in part the aspect of the heterozygosity paradox of the excess of heterozygotes observed in predominantly selfing organisms.

Table 1. The relative heterozygosity $H(1 - F)$ and fixation index at equilibrium when the level of selfing varies between two values in different generations and there are different levels of autocorrelation between selfing in subsequent generations.

	S_1, S_2					
	0.0, 1.0		0.0, 0.2		0.8, 1.0	
Autocorrelation	H	F	H	F	H	F
-1	0.75	0.250	0.950	0.050	0.188	0.812
-0.5	0.714	0.286	0.949	0.051	0.185	0.815
0	0.667	0.333	0.948	0.053	0.182	0.818
0.5	0.600	0.400	0.946	0.054	0.176	0.824
1	0.500	0.500	0.944	0.056	0.166	0.833

Inbreeding and selection: Because inbreeding affects genotypic frequencies, selection in partially inbreeding populations is quite different than in random-mating populations. Both the dynamics of genetic change and the conditions for an equilibrium differ. For example, assume that a new favorable recessive variant arises in a population. Figure 2 gives the rate of incorporation for a random-mating population and two levels of partial selfing (Hedrick, 1983). Notice that even when the selfing level is only 0.25, incorporation of the new variant is much faster than for the random-mating population. The effect occurs because inbreeding reduces the frequency of heterozygotes in which the new variant is hidden by dominance and generates favorable homozygotes that can be selected. One implication of this is that if two populations differ in the rate of inbreeding due to genetic or environmental factors and a new variant is simultaneously introduced in both, then they may greatly differ in allelic frequency at some points in time.

When there is selection at a biallelic locus and assuming the relative fitness of the heterozygote is unity, then depending upon the relative fitness of the homozygotes, $\underline{w}_{11}$ and $\underline{w}_{22}$ for genotypes $\underline{A}_1\underline{A}_1$ and $\underline{A}_2\underline{A}_2$, respectively, various selective outcomes occur. For example, when $\underline{w}_{11}$ and $\underline{w}_{22} < 1$, then there is a stable equilibrium, as seen in the lower left quadrant of Figure 3 for a random-mating population (Hedrick, 1983). However, when there is inbreeding, the

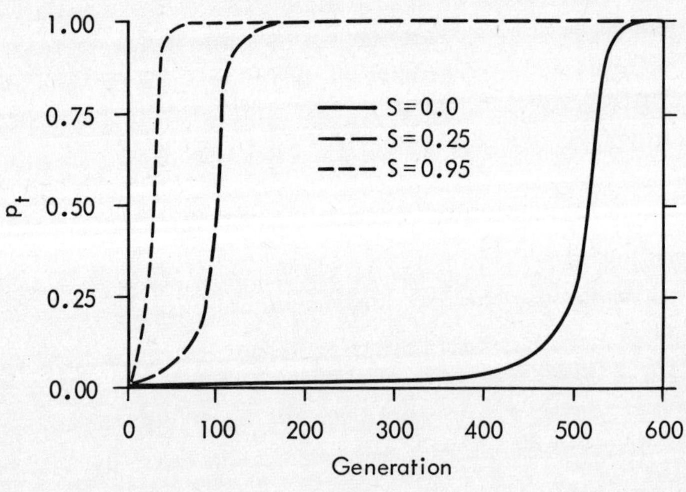

Figure 2. The rate of incorporation of a new recessive variant with three levels of partial self fertilization where $\underline{p}_t$ is the frequency of the favorable $\underline{A}_1$ allele (Hedrick, 1983).

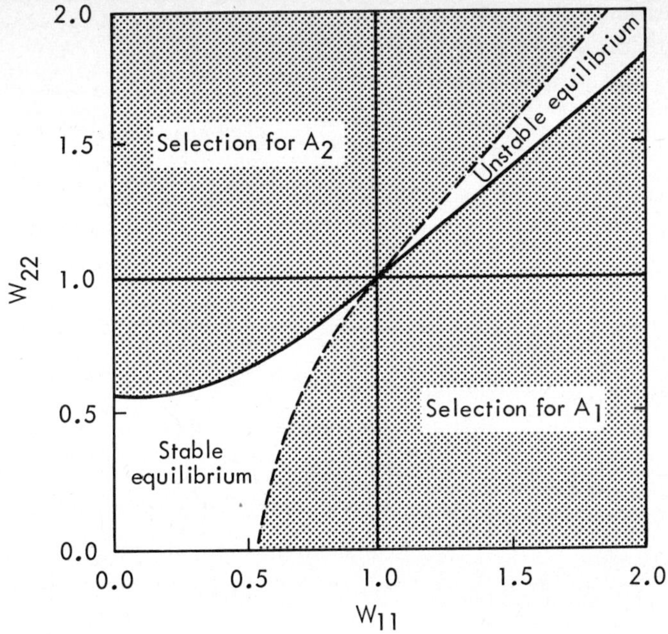

Figure 3. The regions of equilibrium (unshaded) and directional selection when the relative fitnesses at a locus are $\underline{w}_{11}$, 1, and $\underline{w}_{12}$ and there is 0.95 self fertilization (Hedrick, 1983).

size of the regions change and that for a stable equilibrium is reduced. For a polymorphic equilibrium with partial selfing

$$S < \frac{2w_{22}(1 - w_{22})}{w_{11} + w_{22} - 2w_{11}w_{22}} \qquad [8a]$$

and

$$S < \frac{2w_{11}(1 - w_{11})}{w_{11} + w_{22} - 2w_{11}w_{22}} \qquad [8b]$$

(Kimura and Ohta, 1971).

Assuming that a partial full-sib mating model affects equilibrium heterozygosity such that

$$S = \frac{S^*}{2 - S^*} \qquad [9]$$

as derived from the equations above, then the conditions for a stable equilibrium with partial full-sib mating are

$$S^* < \frac{4w_{22}(1 - w_{22})}{w_{11} + 3w_{22} - 2w_{22}^2 - 2w_{11}w_{22}} \qquad [10a]$$

and

$$S^* < \frac{4w_{11}(1 - w_{11})}{w_{11} + 3w_{22} - 2w_{22}^2 - 2w_{11}w_{22}} \quad . \qquad [10b]$$

Likewise if we assume there is both selfing and full-sib mating, then

$$\frac{2S + S^*}{2 - S^*} < \frac{2w_{22}(1 - w_{22})}{w_{11} + w_{22} - 2w_{11}w_{22}} \qquad [11a]$$

and

$$\frac{2S + S^*}{2 - S^*} < \frac{2w_{22}(1 - w_{11})}{w_{11} + w_{22} - 2w_{11}w_{22}} \qquad [11b]$$

for an equilibrium.

For these models, when there is high inbreeding and w_{11} and w_{22} are close to unity, i.e., little selection against the homozygotes, the conditions for stability are quite restrictive so that there can be little asymmetry in the fitness values (see the unshaded region in Figure 3). For example, if $S = 0.95$ (or $S^* = 0.974$ or $S = 0.7$ and $S^* = 0.206$) and $w_{22} = 0.9$, then w_{11} must be between 0.888 and 0.91. One implication of this is that in a highly inbreeding species, slight differences in selection may lead to large differences in allelic frequencies. For example, if again $S = 0.95$ and $w_{22} = 0.9$, then when $w_{11} < 0.888$, A_2 would go to fixation while if $w_{11} > 0.91$, A_1 would go to fixation.

However, for symmetrical or nearly symmetrical selection models, the rate of allelic frequency change is quite slow. For example, Figure 4 gives the change in allelic frequency when $w_{11} = w_{22} = 1.25$, a heterozygous disadvantage, for three levels of partial selfing (Hedrick, 1980a). With high selfing, the allelic frequency only slowly moves away from the unstable equilibrium. This slow response occurs because with high inbreeding, there are few heterozygotes, making selection primarily between the homozygotes. With complete symmetry, there is no difference between the homozygotes and the rate of response is mainly a function of the frequency of the few heterozygotes in the population.

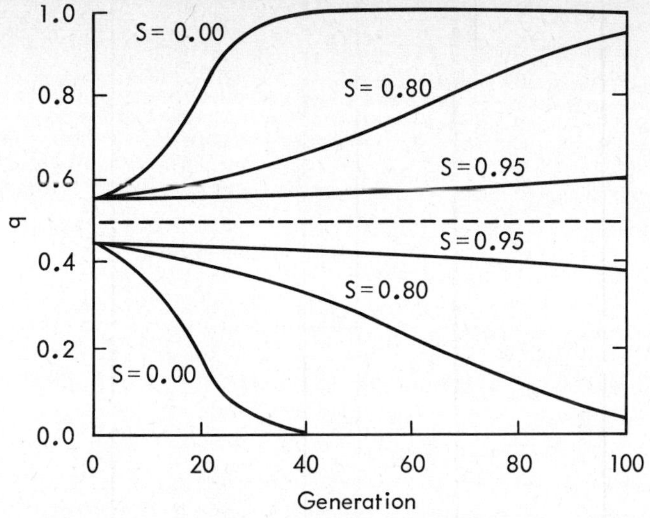

Figure 4. The chance in allelic frequency when the relative fitnesses are 1.25, 1, and 1.25 for three levels of partial selfing (Hedrick, 1980a).

Spatial variation in selection and gene flow: One of the important types of selection models suggested for the maintenance of polymorphism is spatial variation in selection (e.g., Hedrick et al., 1976; Hedrick, 1986). When there is random mating over the area in which the different types of selection occur, then the conditions under which polymorphism may occur are generally quite restrictive (e.g., Maynard Smith and Hoekstra, 1980). However, limited gene flow between the environmental patches can allow a global polymorphism to be maintained that would be lost in a random-mating population (Christiansen, 1975).

For example, if we assume selection favors A_1 in niche 1 and A_2 in niche 2 by equal amounts (there is additivity in both environments and the selective difference between homozygotes is 0.4), m is the proportion of gene flow from one niche to another, and the niches occur in equal proportions, then the equilibrium frequencies within each niche and overall are as given in Figure 5 (Hedrick, 1983). Even with $m = 0.5$, there is an overall polymorphism because of the symmetry in the model.

If we assume that niche 2 is more common and occurs 0.625 of the time, then the equilibria frequencies are as given in Figure 6. In this case, when $m > 0.28$, the overall polymorphism disappears and the population is fixed for A_2. In other words, the patch favoring A_1 disappears because the Δq from gene flow decreasing

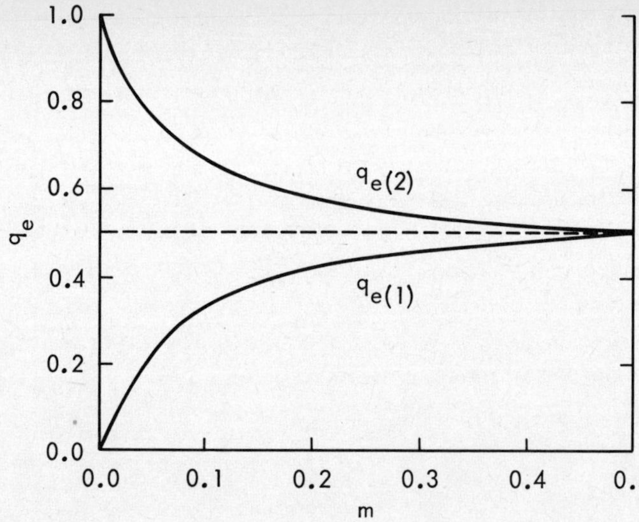

Figure 5. The equilibrium allelic frequency for different levels of gene flow when the relative fitnesses in environment 1 are 1.2, 1, and 0.8 and 0.8, 1, and 1.2 in environment 2 and the environments occur in equal proportions. The solid lines indicate the frequencies in each niche and the broken line is the global equilibrium frequency (Hedrick, 1983).

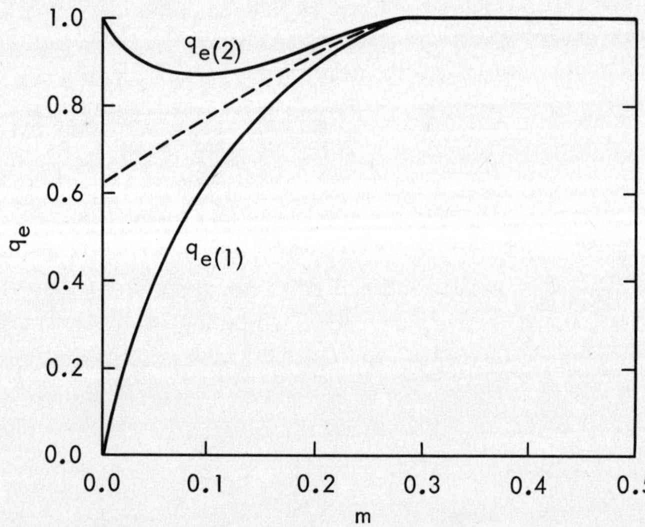

Figure 6. As Fig. 5 except that environment 2 is more common and occurs 0.625 of the time (Hedrick, 1983).

the frequency of A_1 in niche 1 is stronger than the Δq from selection increasing the frequency of A_1.

Spatial variation in selection, gene flow and inbreeding: Let us include inbreeding in the form of partial selfing and specify two types of gene flow as it occurs in plants to further examine this model. First, assume that there is only pollen gene flow from one niche to the other. In this case, gene flow and fertilization occur simultaneously and selection subsequently, say on seedling survival. We can imagine that a certain proportion of flowers produce all selfed progeny while others outcross, some of which results in pollen gene flow. Therefore, recursion equations for the frequency of genotypes A_1A_1, A_1A_2, and A_2A_2 in niche i when there is one other niche j are

$$P_{t+1\cdot i} = (S(P_{t\cdot i} + \tfrac{1}{4}H_{t\cdot i}) + T(p_{t\cdot i}^{\,2}(1 - m) + p_{t\cdot i}\,p_{t\cdot j}\,m))\frac{w_{11\cdot i}}{\bar{w}_i}$$

$$H_{t+1\cdot i} = (\tfrac{1}{2}SH_{t\cdot i} + T(2p_{t\cdot i}\,q_{t\cdot i}(1 - m) + (p_{t\cdot i}\,q_{t\cdot j} + p_{t\cdot j}\,q_{t\cdot i})m))\frac{w_{12\cdot i}}{\bar{w}_i} \qquad [12]$$

$$Q_{t\cdot 1\cdot i} = (S(Q_{t\cdot i} + \tfrac{1}{4}H_{t\cdot i}) + T(q_{t\cdot i}^{\,2}(1 - m) + q_{t\cdot i}\,q_{t\cdot j}\,m))\frac{w_{22\cdot i}}{\bar{w}_i}$$

where the relative fitness of genotypes A_1A_1, A_1A_2, and A_2A_2 in niche i are $w_{11\cdot i}$, $w_{12\cdot i}$, and $w_{22\cdot i}$, respectively. The mean fitness in niche i, $\bar{w}_i$, is the sum of the three equations excluding $\bar{w}_i$.

The second gene flow model assumes only seed gene flow between the niches. In this case, fertilization occurs first, then migration, and finally selection. It is assumed that all pollen stays within the niche. The recursion equations are then

$$P_{t+1\cdot i} = ((1 - m)(S(P_{t\cdot i} + \tfrac{1}{4}H_{t\cdot i}) + Tp_{t\cdot i}^{\,2}) + m(S(P_{t\cdot j} + \tfrac{1}{4}H_{t\cdot j}) + Tp_{t\cdot j}^{\,2}))\frac{w_{11\cdot i}}{\bar{w}_i}$$

$$H_{t+1\cdot i} = ((1 - m)(\tfrac{1}{2}SH_{t\cdot i} + T2p_{t\cdot i}\,q_{t\cdot i}) + m(\tfrac{1}{2}SH_{t\cdot j} + T2p_{t\cdot j}\,q_{t\cdot j}))\frac{w_{12\cdot i}}{\bar{w}_i} \qquad [13]$$

$$Q_{t+1\cdot i} = ((1 - m)(S(Q_{t\cdot i} + \frac{1}{4} H_{t\cdot i}) + Tp_{t\cdot i}^{2})$$

$$+ m(S(Q_{t\cdot j} + \frac{1}{4}H_{t\cdot j}) + Tp_{t\cdot j}^{2}))\frac{w_{12\cdot i}}{\bar{w}_{i}} .$$

Let us compare the effectiveness of these two types of gene flow in maintaining a polymorphism. Figure 7 gives the global equilibrium allelic frequency when the relative fitnesses in environment 1 are 1.1, 1, and 0.9 and 0.8, 1, and 1.2 in environment 2 for different levels of gene flow. It appears that pollen gene flow is much more likely to lead to a polymorphism than seed gene flow, e.g., the polymorphism disappears when $\underline{m}$ = 0.2 for seed gene flow and only when $\underline{m}$ = 0.4 for pollen gene flow. The basis for this difference lies in the simple fact that pollen gene flow is the movement of half a zygote and seed gene flow a whole zygote. When pollen gene flow is rescaled to take this into account, divided by two, the conditions for a polymorphism and the global equilibrium frequency are identical for the two types of gene flow. Furthermore, they are also the same as that assuming Hardy-Weinberg proportions by Christensen (1975). However, the equilibrium allelic frequencies in each niche and genotypic frequencies are slightly different for the two types of gene flow. For example, when $\underline{q}_e$ = 0.75, the

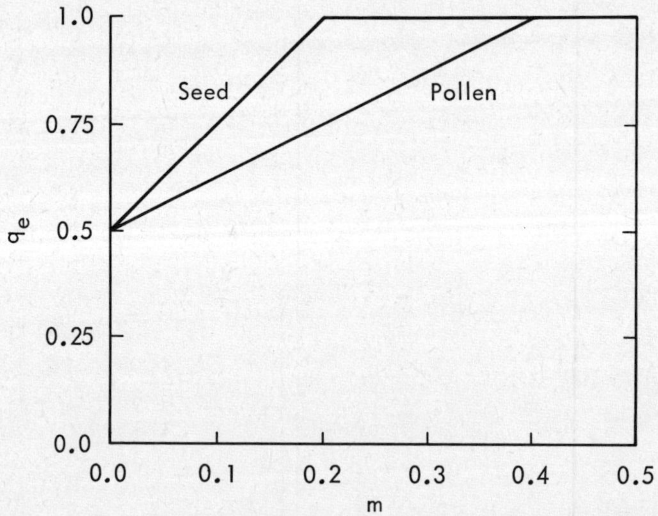

Figure 7. The equilibrium allelic frequency for different levels of gene flow, either by seed or pollen, when the relative fitnesses are 1.1., 1, and 0.9 in environment 1 and 0.8, 1 and 1.2 in environment 2.

fixation indices averaged over both environments are 0.025 and -0.006 for seed and pollen gene flow, respectively. The deficiency of heterozygotes for seed gene flow is due to the Wahlund effect while the excess of heterozygotes for pollen gene flow is caused by the difference in gametic frequencies in the two environments.

How do different levels of selfing affect the potential for a polymorphism? Figure 8 gives the equilibrium values for three levels of pollen gene flow when the relative fitnesses are 1.05, 1, and 0.95 in environment 1 and 0.8, 1, and 1.2 in environment 2. Obviously, higher levels of selfing increase the likelihood of a global polymorphism, basically by causing a reduction in the extent of gene flow. For example, when $\underline{m} = 0.5$, there is no polymorphism possible for these fitness values but when $\underline{S} \geqq 0.6$, a polymorphism does occur. This is in contrast to the heterozygote advantage model in which higher selfing makes a polymorphism less likely. In other words, balancing selection dúe to variation of selection in space seems a more likely mechanism of global polymorphism maintenance in predominantly selfed plants than heterozygote advantage. On the other hand, if gene flow is large and inbreeding low and either the environmental patches differ in size or selection is not symmetrical, an allele that is advantageous in an environment may disappear because favorable selection cannot counteract the effect of gene flow.

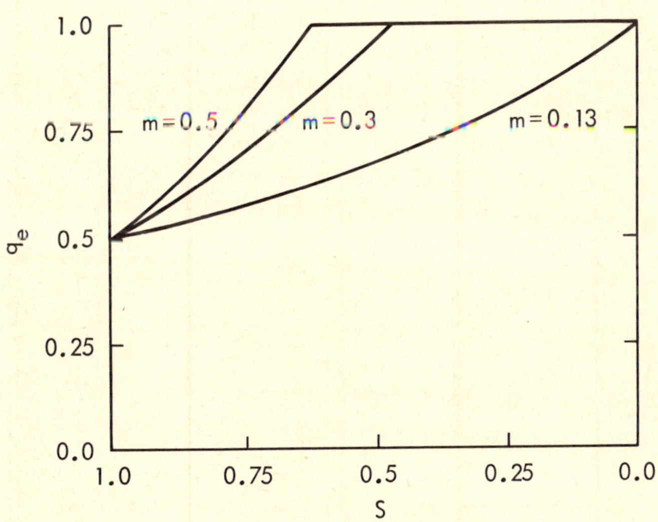

Figure 8. The equilibrium allelic frequency for different levels of self-fertilization and three levels of gene flow when the relative fitnesses are 1.05, 1, and 0.95 in environment 1 and 0.8, 1, and 1.2 in environment 2.

Multiple Loci

Neutrality expectations: The neutrality theory assumes that alleles are selectively equivalent, variants are brought into the population by mutation and eliminated by genetic drift. If a sample of size $2n$ is drawn from a population in neutrality equilibrium and it has k alleles, then it has an expected Hardy-Weinberg homozygosity (Ewens, 1972). For example, if a sample of size 400 has four alleles with frequencies 0.78, 0.16, 0.05, and 0.01 and consequently a homozygosity of 0.637, it is very close to neutrality expectations. Inbreeding should not have any effect on one-locus expectations because Hardy-Weinberg homozygosity is used.

The neutrality model can be extended to two linked loci. In this case, the association between alleles at two loci is a function of the recombination between them as well as the sample size and number of alleles at each locus. For example with very tight linkage, recombination is not able to break down the statistical association that is present when a new variant occurs. Therefore, a given amount of overall gametic disequilibrium (association) is expected in a sample of size $2n$ with k and l alleles at two loci and recombination c between them (Hedrick and Thomson, 1985). In this case, inbreeding should have an analogous effect as linkage, i.e., higher inbreeding should increase the expected disequilibrium in a manner similar to decreased recombination.

One measure that is useful for comparisons between the expected values from neutrality and that for a given sample is the distribution of the standardized gametic disequilibrium, D' (Lewontin, 1964). D' is useful because it is a measure of the association of alleles at two loci, independent of the allelic frequencies at the loci. Figure 9 gives the expected distribution of D' in a sample of size $2n = 200$ with six alleles at each locus (Hedrick and Thomson, 1985). Notice that the distribution is nearly symmetrical with the exception that the -1.0 class is quite large. A value of $D' = -1$ occurs when a particular gametic type is missing from the sample.

Except for a few organisms, such as *Drosophila melanogaster* and humans, the map distance between linked loci is often not precisely known. One exception is for a group of electrophoretic loci in lodgepole pine, four of which are closely linked on one chromosome and two of which are linked on another chromosome. Epperson (1983) surveyed the frequencies of alleles at these loci and the frequencies of two locus gametes in samples of approximately $2n = 400$ in two different populations. The distributions of D' values for the most closely linked loci, each with three alleles, are given in Figure 10. This distribution has the same general pattern as found for neutrality although there is much less disequilibrium. In particular, there are many more values around $D' = 0$ and fewer $D' = -1$ and $D' = 1$ values as compared to neutrality (see Figure 9).

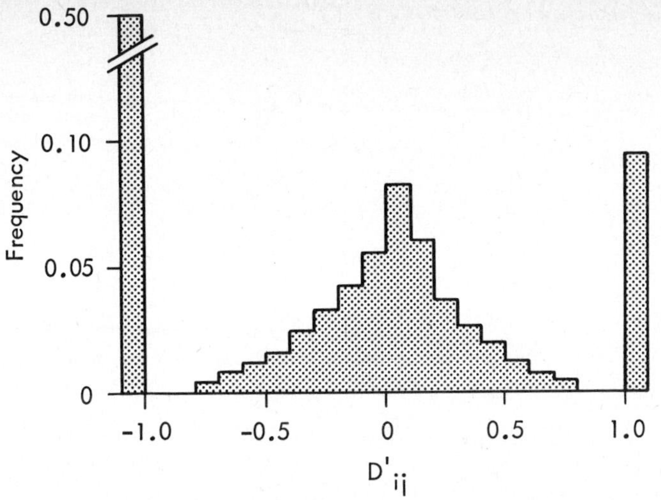

Figure 9. The theoretical distribution of D_{ij}' in a sample of size $2n$ = 200 with six alleles at each locus.

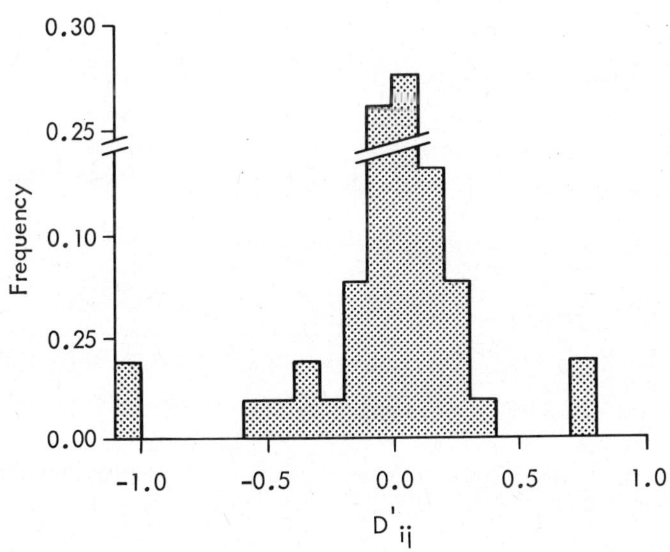

Figure 10. The observed distribution of D_{ij}' in a sample from lodgepole pine.

Selection: Most of the initial two-locus selection studies in partially inbreeding organisms suggested that inbreeding and linkage have similar effects (e.g., Allard et al., 1968). This finding was supported by the fact that inbreeding and linkage retard the rate of decay of the gametic disequilibrium in the same way, i.e., at the equilibrium rate of decay

$$D_{t+1} = \frac{1}{2}\left(\frac{1 + \lambda + S}{2} + \left(\left(\frac{1 + \lambda + S}{2}\right)^2 + 2S\lambda\right)^{\frac{1}{2}}\right)D_t \qquad [14]$$

where $\lambda = 1 - 2\underline{c}$ (Karlin, 1969; Weir et al., 1972). This equivalence is due to a reduction in the production of new gametes from double heterozygotes caused either by lower recombination or fewer double heterozygotes by inbreeding.

However, as shown by Holden (1979) when selection is present, the equivalence is no longer true. As an illustration, let us consider a symmetrical fitness array first investigated by Lewontin and Kojima (1960) (see Table 2). Without inbreeding for there to be a stable two-locus polymorphism with gametic disequilibrium

$$c < \frac{a - 2b + 1}{4}$$

where $\underline{a} - 2\underline{b} + 1$ is a measure of epistasis.

Table 2. A completely symmetrical two-locus fitness array.

	A_1A_1	A_1A_2	A_2A_2
B_1B_1	a	b	a
B_1B_2	b	1	b
B_2B_2	a	b	a

Three examples were discussed by Holden (1979) to illustrate the interaction of selection and inbreeding (Figure 11). In the top example for no inbreeding ($\underline{S} = 0$), only the polymorphism without disequilibrium is present. However, as $\underline{S}$ increases to 0.5 and above, a polymorphism with gametic disequilibrium results. In this case, more inbreeding and tighter recombination have analogous effects.

In the other two examples, this is not true. In the center example of Figure 11 as inbreeding increases, the polymorphism with gametic disequilibrium that is present with no inbreeding disappears, the opposite of the top example. The bottom example presents an even more complex case in which the polymorphism with

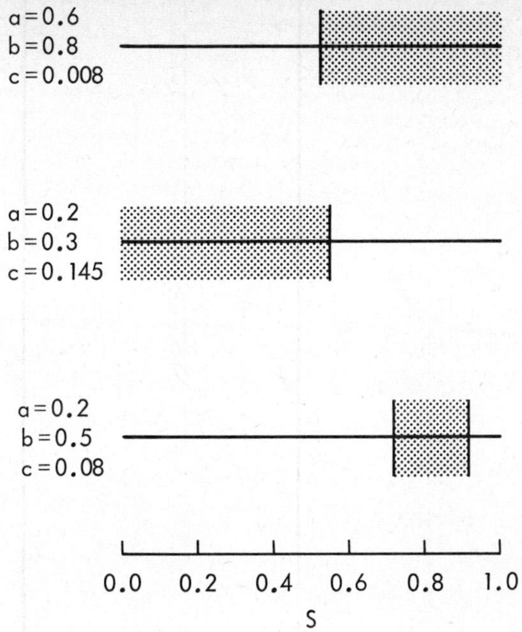

Figure 11. The regions of two locus equilibria with gametic disequilibrium (shaded) for different levels of selfing and three fitness-recombination sets (Holden, 1979).

gametic disequlibrium exists only for an intermediate range of selfing. The basis for these results lies in the complex manner in which inbreeding and selection influence the frequencies of the single- and double-heterozygotes (see Holden, 1979 for a discussion). In other words, gametic disequilibrium can increase or decrease as inbreeding is increased. Furthermore, increased selfing, unlike tighter linkage, may result in a decreased fitness, a result shown by Holden (1979) and expanded by Charlesworth et al. (1979). Note that different populations, having the same selection but slightly different amounts of inbreeding, could have quite different gametic and genotypic frequencies due to those effects.

Genetic hitchhiking: If alleles at a neutral or nearly neutral locus are in gametic disequilibrium with alleles at a selected locus, then the neutral alleles may be carried along, or hitchhike, as selection changes allelic frequencies at the other locus. For hitchhiking to occur there must be initial disequilibrium, a factor that slow the decay of the association, and selective change that occurs before the association disappears (Hedrick, 1980b). In some cases, genetic hitchhiking may increase gametic disequilibrium over time between two neutral loci (Hedrick and

Holden, 1979). Let us consider a particular situation in which the initial disequilibrium is caused by founder effect, inbreeding slows the rate of decay, and selection results from spatial heterogeneity in the environment.

This model was developed to mimic possible changes in a population of wild oats in California (Allard et al., 1972) by Hedrick and Holden (1979). Assume that the area was colonized originally by equal number of genotypes $\underline{A}_1\underline{B}_1\underline{C}_1/\underline{A}_1\underline{B}_1\underline{C}_1$ and $\underline{A}_2\underline{B}_2\underline{C}_2/\underline{A}_2\underline{B}_2\underline{C}_2$ and that the area is divided into two different niches where the fitnesses of the genotypes $\underline{A}_1\underline{A}_1$, $\underline{A}_1\underline{A}_2$, and $\underline{A}_2\underline{A}_2$, at a selected locus unlinked to the other loci are 1.2, 1.0 and 0.8 in one niche, and 0.8, 1.0 and 1.2 in the second niche (these selective differences could be less or spread out over several loci and the same argument made). Assume also that there is limited seed gene flow between niches ($\underline{m}$ = 0.05) and there is 0.98 self-fertilization (Hamrick and Allard, 1972).

Using these values, we obtain the allelic frequencies (the allelic frequency in the other niche is the complement of the one given) and within niche $\underline{D}'$ values given in Figure 12 for a time course of 200 generations. The simulated allelic frequencies are of the same magnitude as those observed at the different sites. For example, the difference in allelic frequency between the two simulated niches is 0.68 after 100 generations, compared to the observed difference between sites of 0.71. The open circles in Figure 12 indicate when between generations 100 and 200 the observed within-niche $\underline{D}'$ value occurs in the simulation for the different map distances (the observed value for $\underline{c}$ = 0.27 is less than the simulated value at

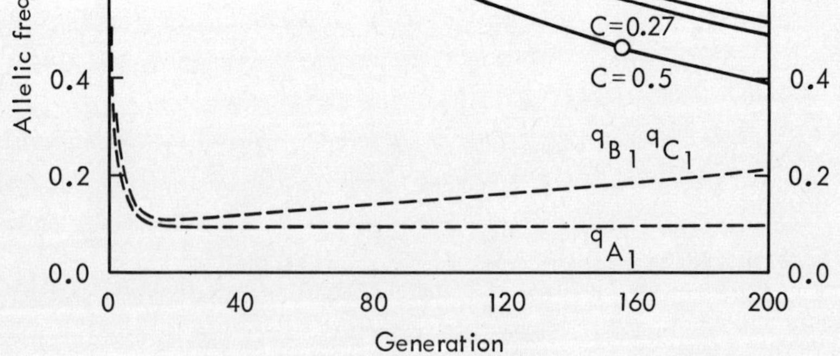

Figure 12. The simulated change in allelic frequencies gametic disequilibrium for one selected locus ($\underline{A}$) and two neutral loci ($\underline{B}$ and $\underline{C}$). The open circles indicate observed values in a wild oat population (Hedrick and Holden, 1979).

generation 200). Of course, greater selection, linkage of the selected locus, reduced gene flow between niches, and a younger population age could make the simulated effects more pronounced.

It is obvious that the observed allelic frequencies and gametic disequilibrium values could be the result of hitchhiking of the allozyme loci on other loci undergoing directional selection. Because there is no temporal information on allelic and gametic frequencies in this population of Avena, it is difficult to determine the cause of the present-day gametic arrays. Overall the data are consistent with a model like that suggested above, which assumes differential selection at a locus other than the allozyme loci examined.

Conclusions

There are a number of situations important in plants that have not been thoroughly explored theoretically. Particularly, when the mating system, here in the form of partial inbreeding, is taken into account, predictions about the effects of other evolutionary factors are often not obvious. In other words, one should be cautious when predicting the effects of more biologically realistic models that simultaneously incorporate several population genetic parameters on the pattern of genetic variation in natural populations. Let me summarize some of the conclusions included in this paper to illustrate.

(1) When inbreeding in addition to partial selfing is examined for a neutral locus, the equilibrium heterozygosity, is lowered. Such inbreeding may be important in explaining part of the heterozygosity paradox, i.e., high outcrossers have a deficiency of heterozygotes.

(2) Temporal autocorrelation of inbreeding can affect the equilibrium heterozygosity in a population. Heterozygosity is highest when the autocorrelation is negative and lowest when the autocorrelation is positive. If there is a negative autocorrelation in inbreeding, this may be of some importance in explaining the higher than expected heterozygosity of predominant selfers.

(3) Partial inbreeding may either increase or decrease the rate of genetic change due to selection. For example, a recessive variant is incorporated more quickly in an inbreeding population while the genetic change for a locus with heterozygous advantage or disadvantage is slowed by inbreeding.

(4) Partial inbreeding may either decrease or increase the conditions for a polymorphism. For example, the conditions for a stable polymorphism by heterozygous advantage are more restrictive for inbreeders while they are less restrictive when there is spatial variation in selection.

(5) Pollen and seed gene flow of the same standardized magnitude have identical effects on the conditions for a genetic polymorphism. However, the

equilibrium genotypic proportions are slightly different, a fact that may be of significance in some cases.

(6) There is an expectation of gametic disequilibrium for two-locus, multiple allele samples from a population at genetic drift-mutation equilibrium. However, data from lodgepole pine exhibit less disequilibrium than neutrality expectations.

(7) Inbreeding and recombination may have very different effects in multilocus selection models. For example, increased inbreeding may reduce the likelihood of a two-locus polymorphism with gametic disequilibrium and reduce fitness while recombination never has these effects.

(8) Genetic hitchhiking can explain the presence of gametic disequilibrium between neutral loci such as in wild oats. In highly selfed plants, the selected locus may be unlinked to the observed loci and still have profound effects on their dynamics.

References

Allard, R. W., G. R. Babble, M. T. Clegg, and A. L. Kahler. 1972. Evidence for coadaptation in Avena barbata. Proc. Nat. Acad. Sci. 69:3043-3048.

Allard, R. W., S. K. Jain, and P. L. Workman. 1968. The genetics of inbreeding populations. Advan. Genet. 14:55-131.

Brown, A. H. D. 1979. Enzyme polymorphism in plant populations. Theoret. Pop. Biol. 15:1-42.

Brown, A. H. D., S. C. H. Barrett, and G. F. Moran. 1985. Mating system estimation in forest trees: Models, methods, and meanings. This volume.

Charlesworth, D., B. Charlesworth and C. Strobeck. 1979. Selection for recombination in partially self-fertilizing populations. Genetics 93:237-244.

Christiansen, F. B. 1975. Hard and soft selection in a subdivided population. Amer. Natur. 109:11-16.

Epperson, B. K. 1983. Multilocus genetic structure of natural populations of lodgepole pines. Ph.D. Thesis, Univ. Calif.-Davis.

Ewens, W. J. 1973. The sampling theory of selectively neutral alleles. Theoret. Pop. Biol. 3:87-112.

Hamrick, J. and A. Schnabel. 1985. Understanding the genetic structure of plant populations: Some old problems and a new approach. This volume.

Hedrick, P. W. 1980a. The establishment of chromosomal variants. Evolution 35:322-332.

Hedrick, P. W. 1980b. Hitchhiking: A comparison of linkage and partial selfing. Genetics 94: 791-808.

Hedrick, P. W. 1983. Genetics of Populations. Jones and Bartlett, Boston, Mass.

Hedrick, P. W. 1985a. Partial inbreeding other than self-fertilization: the effect on equilibrium heterozygosity.

Hedrick, P. W. 1985b. Temporal and spatial variation in self-fertilization: The effect on heterozygosity, polymorphism and gametic disequilibrium.

Hedrick, P. W. 1986. Genetic polymorphism in heterogeneous environments: A decade later. Ann. Rev. Ecol. and Syst. (in press).

Hedrick, P. W., M. E. Ginevan, and E. P. Ewing. 1976. Genetic polymorphism in heterogenous environments. Ann. Rev. Ecol. Syst. 7:1-32.

Hedrick, P. W., and L. Holden. 1979. Hitch-hiking: An alternative to coadaptation for the barley and slender wild oat examples. Heredity 43:79-86.

Hedrick, P. W. and G. Thomson. 1985. A two-locus neutrality test with applications. Genetics (in press).

Holden, L. R. 1979. New properties of the two-locus partial selfing model with selection. Genetics 93:217-236.

Karlin, S. 1969. Equilibrium Behavior of Population Genetics Models with Non-random Mating. Gordon and Breach, London.

Kimura, M. and T. Ohta. 1971. Theoretical Aspects of Population Genetics. Princeton Univ. Press, Princeton, N.J.

Lewontin, R. C. 1964. The interaction of selection and linkage. I. General considerations; heterotic models. Genetics 49:49-67.

Li, C. C. 1976. First Course in Population Genetics. Boxwood Press, Pacific Grove, California.

Maynard, Smith, J., and R. Hoekstra. 1980. Polymorphism in a varied environment: how robust are the models. Genet. Res. 35:45-57.

Nei, M. 1975. Molecular population genetics and evolution. American Elsevier, New York.

Ritland, K. 1984. The effective proportion of self-fertilization with consanguineous matings in inbred populations. Genetics 106:139-152.

Weir, B. S., R. W. Allard, and A. L. Kahler. 1972. Analysis of complex allozyme polymorphisms in a barley population. Genetics 72:505-523.

POLYMORPHIC EQUILIBRIA UNDER INBREEDING EFFECTS AND SELECTION ON COMPONENTS OF REPRODUCTION

Martin Ziehe

ABSTRACT

From previous investigations on genotypic equilibria existence, we know that a diallelic polymorphic overdominance equilibrium cannot simultaneously show a high asymmetry in homozygotic fitness disadvantage and a strong deficit in heterozygote frequency relative to Hardy-Weinberg proportions. Since on the other hand sexually asymmetrical selection during reproduction is able to compensate a reduction in heterozygote frequency originating from inbreeding effects, the question arises as to whether the above property becomes meaningless in the presence of sexually asymmetrical selection. To answer this, polymorphic equilibria are calculated and also characterized for a multiallelic model of fertility resource allocation and partial self-fertilization where the fertilities are allowed to show extreme sexual asymmetry.

It turns out that for this model two types of equilibria can occur. For equilibria of the first type the above mentioned property can serve to investigate their existence. For equilibria of the second type all genotypic fitnesses are always identical, so that the above property is trivially fulfilled and does not provide any further insights. Moreover, the latter type of equilibrium directly reflects the influence of sexual asymmetry.

As a consequence, it is confirmed that sexual asymmetry represents an efficient mechanism for the establishment of additional polymorphic equilibria even in the presence of strong inbreeding effects, and it is suggested that protectedness of polymorphisms can be classified according to the underlying mechanism of either symmetry in homozygotic disadvantage or sexual asymmetry.

1. PREVIOUS RESULTS ABOUT EQUILIBRIUM EXISTENCE

This contribution is devoted to the characterization and classification of polymorphic genotypic equilibria, coupled with a search for conditions implying the non-existence of particular equilibrium types. To begin with, it should be emphasized that non-existence of polymorphic equilibria describes a situation which usually coincides with a loss of genetic multiplicity or even convergence to a fixation state. Non-existence generally does not, however, guarantee fixation, just as existence of a single polymorphic equilibrium, even if it is stable, does not automatically yield a protected polymorphism. Nevertheless, theoretical investigations into equilibrium existence form an important step for predicting the long-term behavior of genotypic structure dynamics and may serve to check the validity of or to falsify hypotheses about evolutionary components: A population observed to be within a polymorphic genotypic equilibrium would falsify model assumptions which rule out the existence of any polymorphic equilibrium. On the other hand, if the model parameters can be estimated for an actual population, and the parameter values are inserted into the model, knowledge of whether these values allow for a polymorphic equilibrium within the frame set by this model can help in predicting whether or not the population is in a polymorphic equilibrium. This latter aspect has particular relevance in forestry, since, due to the longevity of most trees, equilibrium situations cannot easily be observed.

The results for the classical 2-allele selection model with mixed mating may serve to elucidate some of the basic underlying ideas to be discussed later in this paper and hence will be briefly mentioned here. For sexually symmetrical viability selection with purely random mating, it is well known that a viability advantage of heterozygotes is sufficient to guarantee the existence of a polymorphic genotypic equilibrium. Moreover, this polymorphic equilibrium turns out to be globally attractive and thus will be approached from any initially polymorphic genotypic structure. If random mating is replaced by a mixture of random mating and partial self-fertilization, this overdominance criterion no longer remains valid. If the asymmetry between homozygotic viability disadvantage and the proportion of self-fertilized ovules are both sufficiently large, then no polymorphic equilibrium exists (compare Hayman, 1953, Kimura and Ohta, 1971, Brown, 1979 or Hedrick, 1985), and computer simulations show convergence to fixation of one allele. Thus in this case an increase in self-fertilization due, for example, to environmental changes may cause disappearance of a stable viability polymorphism and therefore a loss of genetic variation in the long run.

Preconditions which turn out to be basically responsible for this observation are discussed in a more general context in a previous paper (Ziehe, 1984) and will be briefly recalled here. Consider a hermaphrodite or monoecious plant population reproducing in non-overlapping generations. The population size is assumed to be effectively infinite, so that drift effects can be ignored. All subsequent genetic considerations refer to a single autosomal gene locus with alleles A_1, A_2, ... and unordered genotypes A_1A_1, A_1A_2, A_2A_2,... . The fitness w_{kl} of a genotype A_kA_l describes the average number of gametes individuals of genotype A_kA_l effectively contribute to the offspring generation. Census occurs at the zygotic stage. Let P_{kl} denote the relative frequency of the genotype A_kA_l, p_k the relative frequency of the allele A_k. A parameter or frequency within a genotypic equilibrium will be indicated by an asterisk. Then the main result yields an inequality which must hold within a polymorphic genotypic equilibrium. This basic inequality is illustrated in Figure 1. It formally corresponds to

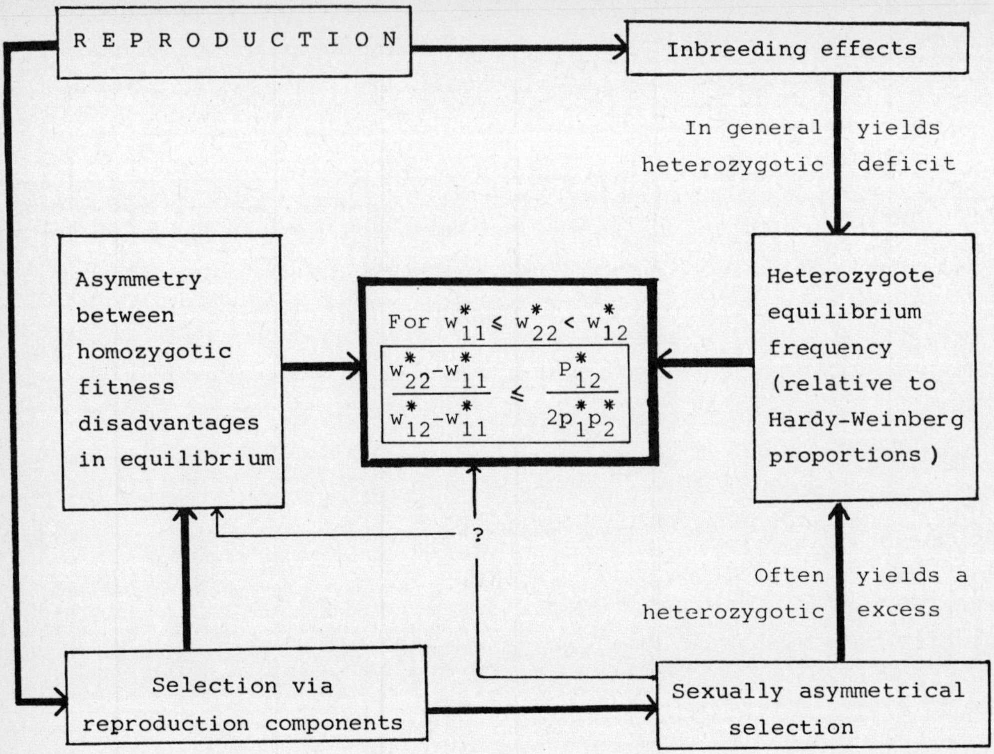

Figure 1: An inequality realized within a diallelic overdominance equilibrium

that derived earlier by Li (1955) for the classical viability selection model. Nevertheless, as demonstrated previously, the inequality is also applicable to selectionary effects originating from reproduction (such as genotypically differential ovule selfing rates or positive assortative mating) and may be interpreted in the following manner: A diallelic polymorphic overdominance equilibrium cannot simultaneously show a high asymmetry in homozygotic fitness disadvantage and a low heterozygote frequency relative to Hardy-Weinberg proportions.

The above result often provides a procedure for determining parameter regions which a priori exclude the existence of polymorphic genotypic equilibria (Ziehe, 1984).

We now turn to an aspect which often has been ignored in selection discussions. If selection acts via reproduction components, sexually asymmetrical selectionary effects generally must be taken into account. This kind of asymmetry can be introduced, for example, by genotypically non-symmetrically varying ovule and pollen production.

Moreover, the special case of sexual symmetry requires that an advantage in ovule production must always be coupled with a corresponding advantage in pollen production and thus seems to be more the exception than the rule.

For certain tree populations, this kind of sexual asymmetry in fertilities has already been observed (see e.g. Ross, 1984, and Ross, 1985, for Pinus sylvestris L.). Investigations of isozymes in pine seeds using electrophoretic methods also indicate that the realized contributions of pine seed orchard clones to the seed crop can vary sexually asymmetrically. Figure 2 illustrates previous results for which it turns out that contributions via pollen lying above the population average are always coupled with below average contributions via ovules and vice versa.

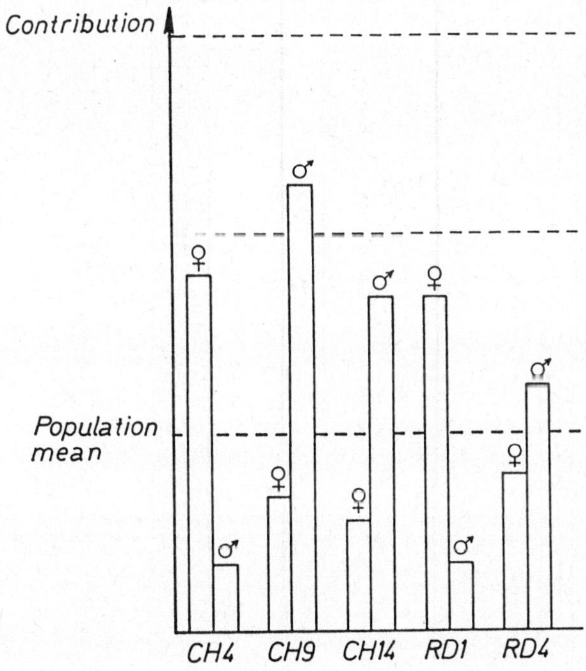

Figure 2: Estimated female (♀) and male (♂) contributions of particular pine seed orchard clones to the seed crop of 1976 (from Ziehe and Hattemer, 1985 as in Müller-Starck and Ziehe, 1984)

It has already been described that sexual asymmetry is able to produce a heterozygotic excess relative to Hardy-Weinberg proportions. As an extreme example, imagine a population in which A_1A_1 genotypes produce only male gametes, A_2A_2 genotypes only female gametes and heterozygotes are not present. Then the offspring consists only of heterozygotes. As might be expected from this, sexual asymmetry is able to compensate a reduction in heterozygote frequency relative to Hardy-Weinberg proportions which is caused by inbreeding effects (Ziehe, 1983). An additional numerical example will be presented later (Table 1, Example 4).

Thus the question arises as to whether the above described mechanism for ruling out the existence of polymorphic equilibria remains useful in that it also functions in the presence of sexually asymmetrical selectionary effects. Earlier investigations on differential pollen production under mixed random mating and self-fertilization show that the above procedure still works (Ziehe, 1982). Therefore, in the following section the model of fertility resource allocation suggested by Ross and Gregorius (1983) will be considered. The model assumptions allow for an essentially larger amount of sexual asymmetry in gamete production than differential pollen production. Hence, the investigations are expected to deliver some new insights with respect to the above question.

2. POLYMORPHIC EQUILIBRIA UNDER FERTILITY RESOURCE ALLOCATION AND PARTIAL SELF-FERTILIZATION

a. Model assumptions and immediate consequences :

Fertility resource allocations (as defined in Ross and Gregorius, 1983): Let a genotype A_kA_l produce an average number ϕ_{kl} of ovules and μ_{kl} of pollen grains. It is assumed that for each genotype a genotypically constant amount of resources is available from which the production of a single ovule consumes the fraction $r^{♀}$, that of a single pollen grain the fraction $r^{♂}$ of these resources. Thus ϕ_{kl} can vary only between 0 and $(r^{♀})^{-1}$ and the remaining resources of A_kA_l suffice to produce on average

$$\mu_{kl} = (1 - \phi_{kl}\, r^{♀})(r^{♂})^{-1}$$

pollen grains. It is evident that a genotypical advantage in ovule production is always coupled with a disadvantage in pollen production and vice versa, so that these model assumptions imply a high degree of sexual asymmetry in fertilities.

The proportion of resources invested into ovule production equals $\phi_{kl}r^{♀}$ for the genotype A_kA_l and subsequently will be abbreviated by R_{kl} . $\bar{R}$ describes the corresponding population mean.

Partial self-fertilization: For individuals of genotype A_kA_l, a mean number $\sigma\,\phi_{kl}$ of ovules is assumed to be reserved for prior fertilization by the individual itself. All remaining ovules will be fertilized at random by pollen from the pollen cloud which consists of all pollen grains not needed for prior self-fertilization. For most plant populations $r^{♂} \ll r^{♀}$ holds, so that the number of pollen used for self-fertilization is negligible as compared to the total number of pollen produced. Therefore it is supposed that the contribution of a particular genotype to the pollen cloud is sufficiently well described by its total pollen production.

Fitness values: Fitnesses of genotypes are frequency dependent and read

$$w_{kl} = (r^{♀})^{-1}\left((1+\sigma)R_{kl} + (1-\sigma)\frac{\bar{R}(1-R_{kl})}{1-\bar{R}} \right).$$

The recurrence equations: It is useful to write the recurrence equations uniquely in terms of the R_{kl}'s . For the allelic frequencies it reads

$$p_k' = \frac{1}{2}\,p_k\left((1+\sigma)\frac{\bar{R}_k}{\bar{R}} + (1-\sigma)\frac{(1-\bar{R}_k)}{(1-\bar{R})} \right)$$

for all k, where the prime indicates the subsequent generation and $\bar{R}_k$ for k=1,2,... denotes the mean allelic proportion of investment into ovule production, i.e.

$$\bar{R}_k = \left(P_{kk}R_{kk} + 0.5\sum_{l\,l\neq k} P_{kl}R_{kl} \right) / \,p_k.$$

For heterozygotes A_kA_l with $k\neq l$ we obtain

$$P_{kl}' = \frac{1}{2}\,\sigma P_{kl}\frac{R_{kl}}{\bar{R}} + (1-\sigma)p_kp_l\frac{\bar{R}_k(1-\bar{R}_l)+\bar{R}_l(1-\bar{R}_k)}{\bar{R}(1-\bar{R})}.$$

Classification of equilibria: It is assumed that the R_{kl}'s are not equal for all genotypes (or else no fertility selection occurs) and that $0<\bar{R}<1$ (or else no reproduction occurs). Using the above allelic recurrence equation we get

$$p_k' - p_k = 0.5\,p_k\,(\bar{R}_k-\bar{R})(1+\sigma-2\bar{R})/(\bar{R}(1-\bar{R})).$$

Thus for a polymorphic genotypic structure with $0<p_k<1$ for each k, the allelic frequencies remain unchanged if and only if at least one of the following conditions is fulfilled :

(i) $\bar{R}_k = \bar{R}$ for all k

(ii) $\bar{R} = (1+\sigma)/2$.

In particular polymorphic genotypic equilibria show allelic constancy and thus also fulfil at least one of the above conditions. Within the subsequent context the equilibria will be classified as follows :

Type 1 equilibria are those for which $\bar{R}_k^* = \bar{R}^*$ for each k,

Type 2 equilibria are those for which $\bar{R}^* = (1+\sigma)/2$.

An equilibrium consequently belongs to both types if $\bar{R}_k^* = \bar{R}^* = (1+\sigma)/2$.

b. Equilibria of type 1:

For an equilibrium of type 1, a comparison with the corresponding Hardy-Weinberg proportions immediately yields

$$P_{kl}^*/(2p_k^*p_l^*) = (1-\sigma)/(1-\sigma R_{kl}/(2\bar{R}^*)) .$$

A direct consequence is that, in the absence of prior self-fertilization (i.e. $\sigma=0$), type 1 equilibria always represent Hardy-Weinberg proportions. Moreover, for arbitrary σ, type 1 equilibria possess the same structural and locational properties as the equilibria for the classical selection model (with sexually symmetrical differential viabilities) and partial self-fertilization provided the viabilities are replaced by the resource parameters R_{kl} and the zygotic stage is considered. Thus for equilibria of type 1, Weir's (1970) method for numerical calculation of the polymorphic equilibrium structures is fully applicable.

The diallelic case: For the diallelic case, $w_{11}^* = w_{12}^*$ yields either $\bar{R}^*=(1+\sigma)/2$, which is a characteristic of type 2 equilibria, or $R_{11} = R_{12}$, which represents the dominance case and is discussed in Ross and Gregorius (1983). Hence we subsequently assume $w_{11}^* \neq w_{12}^*$.

The quotient $(w_{12}-w_{22})/(w_{12}-w_{11})$ is generally used as a measure of fitness asymmetry between homozygotes (see above) and here reduces to $(R_{12}-R_{22})/(R_{12}-R_{11})$ Thus within a type 1 equilibrium, the asymmetry in resource allocation parameters is completely transferred into the corresponding fitness asymmetry.

Since polymorphic genotypic equilibria always show fitness identity for all genotypes, overdominance, or underdominance (Gregorius, 1981), the relations

$$R_{11} < R_{12} < R_{22} \text{ or } R_{22} < R_{12} < R_{11}$$

never lead to polymorphic type 1 equilibria.

From condition (i) it immediately follows that polymorphic type 1 equilibria must always be located on so-called γ-curves (illustrated in Gregorius, 1982 and in Ziehe, 1982), where γ is calculated by

$$\gamma = (R_{12}-R_{22})/(R_{12}-R_{11})$$

and measures the asymmetry between homozygote resource parameters. For $\sigma=0$, a type 1 equilibrium, if it exist at all, lies at the intersection of the corresponding γ-curve and the Hardy-Weinberg parabola. The equilibrium allele frequency of A_1

reads

$$p_1^* = (R_{12}-R_{22})/(2R_{12}-R_{11}-R_{22}) .$$

For arbitrary σ , diallelic type 1 equilibria never exceed the corresponding Hardy-Weinberg proportions. (Proof: Assume for a type 1 equilibrium $P_{12}^* > 2p_1^*p_2^*$ and, without loss of generality, $p_1^* \geq 0.5$. Then $\bar{R}^* = R_2^* \geq p_1^* R_{12} \geq 0.5 \; R_{12}$, which contradicts $(1-\sigma)/(1-0.5\,\sigma R_{12}/\bar{R}^*) > 1$ and thus the assumption $P_{12}^* > 2p_1^*p_2^*$. Therefore $P_{12}^* \leq 2p_1^*p_2^*$ always holds for type 1 equilibria.) An increasing proportion σ of self-fertilized ovules shifts the equilibrium downwards along the γ-curve until fixation is reached. A method for obtaining an idea about the explicit equilibrium location is presented in Ziehe (1982).

The equilibrium exclusion procedure as described above remains completely valid for type 1 equilibria : Both a high amount of asymmetry between homozygotic fitness disadvantages as measured by γ , where γ depends on the R_{kl}'s, and a low frequency of heterozygotes due to partial self-fertilization cannot be realized simultaneously within a type 1 equilibrium. Corresponding parameters of fertility resource allocation and partial self-fertilization may a priori rule out the existence of type 1 equilibria. Of course, for the homozygote symmetry case $R_{11} = R_{22} \neq R_{12}$, a polymorphic type 1 equilibrium always exists.

c. Equilibria of type 2:

Equilibria of type 2 show $\bar{R}^* = (1+\sigma)/2$, which implies

$$w_{kl}^* = (1+\sigma)(r^{♀})^{-1}$$

for each genotype A_kA_l. Thus all genotypic fitnesses are identical and thus equal to the mean population fitness within this type of equilibrium. As an immediate consequence, no asymmetry between homozygotic fitness disadvantages or advantages can occur, and the above equilibrium exclusion procedure does not work.

Note that genotypic fitnesses within type 2 equilibria do not explicitly depend on the allocation parameters R_{kl}.

For $\sigma = 0$, we get $w_{kl}^* = \bar{w}^* = (r^{♀})^{-1}$ .This equals half of the upper bound for the population fitness under identical available resouces in terms of $r^{♀}$, which can approximately be reached by a population with nearly all individuals investing all their resources into ovule production. An increase in the proportion of self-fertilized ovules obviously increases the population fitness within type 2 equilibria.

A type 2 equilibrium directly reflects the influence of sexual asymmetrical selectionary effects: By assumption the amount of ovule production differs between the genotypes, but fitnesses turn out to be identical. Thus an advantage of a certain genotype with respect to its contributions via ovules must be counterbalanced by a

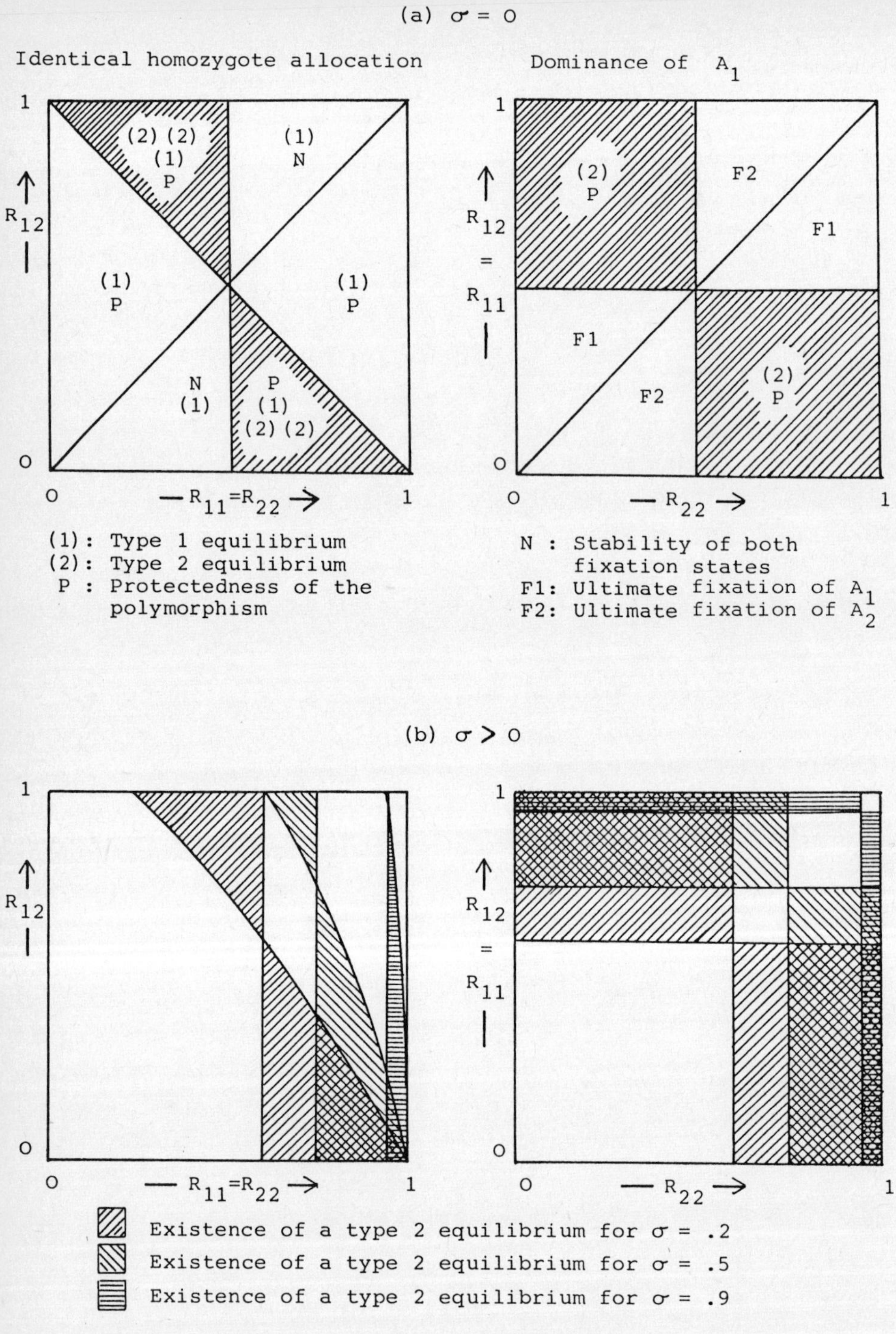

Figure 3: Parameter regions of type 2 equilibria

corresponding disadvantage in reproduction efficiency via pollen. The latter component turns out to be frequency dependent due to pollen competition for cross-fertilization. However, type 2 equilibrium frequencies allow for the exactly counterbalancing effect.

Type 2 equilibria show $\bar{R}^* = (1+\sigma)/2$ and thus clearly cannot exist if

$$\text{either} \quad \max R_{kl} < (1+\sigma)/2$$
$$\text{or} \quad \min R_{kl} > (1+\sigma)/2.$$

For $\sigma = 0$, this requires that within a type 2 equilibrium it is not admissible that all genotypes invest more resource units into ovule than into pollen production or vice versa. For $\sigma > 0$, the genotypic parameters R_{kl} must vary around $(1+\sigma)/2$ in order to allow for this type of equilibria. Nevertheless, this condition is not sufficient to guarantee its existence (compare Fig. 3).

A further characteristic for type 2 equilibria concerns the mean functional sex of the population. The functional sex F_{kl} of a genotype A_kA_l is defined as that proportion of its fitness which is contributed to the offspring generation via ovules. It is given by

$$F_{kl} = R_{kl}/(1+\sigma) .$$

Thus within a type 2 equilibrium the population mean $\bar{F}^*$,

$$\bar{F}^* = \sum P^*_{kl} F_{kl} ,$$

exactly equals 0.5, which corresponds to the functional sex of an ideal bisexual type. This observation does not necessarily hold within type 1 equilibria, even though deviations of $\bar{F}^*$ from 0.5 are generally very small (compare the numerical examples). Figure 3 illustrates the diallelic situation for the extreme cases of identical homozygote resource allocation parameters (left half) and dominance with respect to resource allocation (see also Ross and Gregorius, 1983) (right half). For $\sigma = 0$, the stability properties of equilibria are taken from Bodmer (1963), where for this special case of no prior self-fertilization an analytically equivalent model is investigated.

Fig. 3 shows that an increasing σ shifts the region of those parameters which admit equilibria of type 2, thereby causing the corresponding domain of parameters to become smaller. Thus an increasing proportion σ of self-fertilized ovules may also cause disappearance of type 2 equilibria. Nevertheless, even the case of dominance, which represent an extreme asymmetry in homozygotic resource investment, permits the existence of type 2 equilibria (as already obtained by Ross and Gregorius, 1983).

d. Numerical examples:

Table 1 contains several numerical examples for genotypic structure development. Instead of fitnesses w_{kl}, which depend on the unspecified amount of fertility resources, Table 1 presents values for $r^{♀} w_{kl}$, which measures fitness in terms of resource cost for

Table 1: Examples for the development of genotypic structures

Except example 1a the initial genotypic structure always reads $P_{11}= .1$, $P_{12}= .2$ and $P_{22}= .7$.

Example 1 : $R_{11}= R_{22}= .45$, $R_{12}= .7$ and $\sigma = .1$

Generation	Genotypic structure P_{11}	P_{12}	P_{22}	Fitnesses $r^{♀}w_{11}$	$r^{♀}w_{12}$	$r^{♀}\bar{w}$	$\bar{F}$	I	Equ. type
1	.051	.304	.645	.990	1.040	1.005	.521	.938	
10	.055	.329	.615	1.058	1.077	1.064	.500	.960	
100	.087	.389	.524	1.093	1.096	1.094	.500	.961	
∞	.095	.400	.505	1.100	1.100	1.100	.500	.961	(2)

Example 1a : For initially $p_1= p_2= .5$

Generation	P_{11}	P_{12}	P_{22}	$r^{♀}w_{11}$	$r^{♀}w_{12}$	$r^{♀}\bar{w}$	$\bar{F}$	I	Equ. type
∞	.260	.479	.260	1.151	1.128	1.140	.501	.959	(1)

Example 2 : $R_{11}= R_{22}= .45$, $R_{12}= .7$ and $\sigma = .2$

Generation	P_{11}	P_{12}	P_{22}	$r^{♀}w_{11}$	$r^{♀}w_{12}$	$r^{♀}\bar{w}$	$\bar{F}$	I	Equ. type
1	.063	.285	.651	.980	1.080	1.009	.513	.873	
10	.086	.349	.565	1.048	1.117	1.072	.498	.907	
100	.229	.453	.318	1.107	1.149	1.126	.498	.913	
∞	.272	.457	.272	1.110	1.151	1.128	.498	.913	(1)

Example 3 : $R_{11}= R_{22}= .45$, $R_{12}= .6$ and $\sigma = .1$

Generation	P_{11}	P_{12}	P_{22}	$r^{♀}w_{11}$	$r^{♀}w_{12}$	$r^{♀}\bar{w}$	$\bar{F}$	I	Equ. type
1	.052	.302	.647	.952	.992	.964	.513	.935	
10	.060	.336	.604	.990	1.020	1.000	.499	.953	
100	.158	.452	.391	1.026	1.046	1.035	.499	.955	
∞	.261	.477	.261	1.035	1.053	1.043	.499	.955	(1)

Example 4 : $R_{11}= .45$, $R_{12}= .6$, $R_{22}= .7$ and $\sigma = .1$

Generation	P_{11}	P_{12}	P_{22}	$r^{♀}w_{11}$	$r^{♀}w_{12}$	$r^{♀}\bar{w}$	$\bar{F}$	I	Equ. type
∞	.065	.394	.541	1.100	1.100	1.100	.500	1.020	(2)

a single ovule. For examples 1 to 3 we always have $w_{11} = w_{22}$. Therefore the column for $\overset{\female}{r}\, w_{22}$ is omitted. Additionally the panmictic index I, $I = P_{12}/(2p_1p_2)$, is calculated, which measures the relationship between actual heterozygote frequency and the heterozygotic frequency expected under corresponding Hardy-Weinberg proportions.

Example 1 illustrates convergence from the initial genotypic structure to a type 2 equilibrium. For this particular choice of resource parameters, two type 2 equilibria and a single type 1 equilibrium exist. The type 1 equilibrium is unstable (and attractive only for $p_i = 0.5$) but delivers a higher fitness for each of the genotypes than the stable type 2 equilibria. As expected from the analytical investigations, each of the genotypes shows a type 2 equilibrium fitness of $(1+\sigma)(\overset{\female}{r})^{-1} = 1.1\ (\overset{\female}{r})^{-1}$.

For example 2, the proportion of self-fertilized ovules is increased to 0.2 such that type 2 equilibria no longer exist (compare Figure 3). Convergence occurs to the single type 1 equilibrium, which shows a lower population fitness than the type 1 equilibrium of example 1.

Example 3 differs from example 1 by its reduced ovule investment of the heterozygote. Here again, no type 2 equilibrium can exist, and again convergence occurs to a single type 1 equilibrium.

Example 4 represents a case of intermediate ovule resource investment ($R_{11} < R_{12} < R_{22}$) in which no type 1 equilibrium can exist. It is remarkable that this equilibrium shows a heterozygotic excess relative to Hardy-Weinberg proportions, even though 10% of the ovules are self-fertilized. As mentioned above, this observation suffices to deduce the other type 2 equilibrium characteristics (like $\overset{\female}{r}\, w^*_{kl} = 1+\sigma = 1.1$ or $\bar{F}^* = 0.5$).

3. SUMMARIZING AND CONCLUDING REMARKS

The classical (sexually symmetrical viability) selection together with mixed random mating and self-fertilization can yield an overdominance situation, in which no polymorphic genotypic equilibrium exists. This particular case occurs if the asymmetry between homozygotic viability disadvantages and the proportion of self-fertilized ovules are both extremely large. The mechanism leading to this situation carries over to a generalized model with arbitrary (and thus also reproductive) selection coupled with inbreeding effects: A polymorphic genotypic equilibrium cannot simultaneously exhibit a low heterozygotic frequency and an extreme asymmetry in homozygotic fitness disadvantage, where fitness measures the contribution of gametes which become incorporated into zygotes. For several models of selection via reproduction (as, for example, genotypically differential positive assortative mating), this has been applied to determine combinations of reproduction parameters which a priori exclude the existence of a polymorphic genotypic equilibrium.

The question arises as to whether sexually extremely asymmetrical selection during reproduction, which is capable of counterbalancing inbreeding effects with respect to the frequency of heterozygotes, may also favor the existence of a polymorphic equilibrium. To answer this question, a major part of this paper is devoted to equilibrium investigations for a particular model of fertility resource allocation combined with mixed self-fertilization and random gametic fusion. This model includes an extreme amount of sexual asymmetry ("antisymmetry") in gamete production as well as an inbreeding effect via partial self-fertilization.

We finally obtain two types of equilibria for this particular model of fertility resource allocation and mixed mating.

Type 1 equilibria obey the condition described above: A low heterozygotic frequency and extreme asymmetry in homozygotic fitness disadvantage (or, analoguously, advantage) cannot be realized simultaneously, where the fitness asymmetry can be conveniently measured by a parameter depending uniquely on genotypical resource investment into ovule production. Moreover, type 1 equilibria can be characterized and calculated in the same way as equilibria for the classical viability selection model with mixed mating. Thus, as a basic consequence, this type of equilibrium shows no typical influence originating from sexually asymmetrical effects.

This result accords well with intuitive expectations if one considers the fact that the conditions for type 1 equilibria reflect the situation where all sex-specific allelic fitnesses are identical. Identity of all allelic fitnesses is in turn characteristic of equilibria in models assuming no sex differences, and these models are well known to have the above mentioned properties.

In contrast to type 1 equilibria, the second type of equilibrium (type 2) does reflect the influence of sexually asymmetrical effects. Since for this type of equilibrium, genotypical fitnesses are identical but contributions via ovules to the offspring differ, an advantage in female contribution must always be combined with a corresponding disadvantage in male contribution, so that the sum of both is the same for each genotype. As the results show, this type of equilibrium may even exist in cases where both the proportion of self-fertilized ovules and the asymmetry in resource parameters are so large that type 1 equilibria no longer occur.

From additional computer simulations, we know that deviations from the model assumption of genotypically constant fertility resources modify the list of characteristics for particular equilibria. Nevertheless, for reproduction with a strong inbreeding effect, sexually asymmetrical selection turns out to represent an efficient mechanism for the establishment of further polymorphic equilibria.

An important aspect worthy of further investigation might certainly be the classification of parameter regions which lead to a protected polymorphism according to the underlying mechanism of either symmetry in homozygotic disadvantage or sexual asymmetry. At least for the above model of fertility resource allocation together with mixed mating, this classification should be possible and meaningful.

Acknowledgement:

The author wishes to thank E. Gillet for her help in preparing the manuscript. This work was supported by a grant from the Deutsche Forschungsgemeinschaft, Bonn-Bad Godesberg.

REFERENCES

Bodmer, W. F. 1965: Differential fertility in population genetics models. Genetics 51, 411-424

Brown,A.H.D. 1979. Enzyme polymorphism in plant populations. Theoretical Population Biology 15, 1-42

Gregorius, H.-R. 1981. Realized genotypic fitnesses at equilibrium in the deterministic selection theory of a diallelic locus. Göttingen Research Notes in Forest Genetics No. 4, Göttingen

Gregorius, H.-R. 1982. The relationship between genic and genotypic fitnesses in diploid populations. Evolutionary Theory 6, 143 162

Hayman, I. 1953. Mixed selfing and random mating when homozygotes are at a disadvantage. Heredity 7, 185-192

Hedrick, P.W. 1985. Inbreeding and selection in natural populations. This volume.

Li, C.C. 1955. The stability of an equilibrium and the average of a population.American Naturalist 89,281-296

Kimura, M. and T. Ohta 1971. Theoretical aspects of population genetics. Princeton University Press, Princeton, N.J.

Müller-Starck, G. 1985. Reproductive efficiency of seed orchard clones. This volume.

Müller-Starck, G. and M. Ziehe 1984. Reproductive efficiency in conifer seed orchards. 3.Female and male fitnesses of individual clones realized in seeds of Pinus sylvestris L. Theoretical and Applied Genetics 69, 173-177

Ross, M.D. 1984. Frequency-dependent selection in hermaphrodites: the rule rather than the exception. Biol. J. Linn. Soc. 23, 145-155

Ross, M.D. 1985. Evolution of outbreeding systems. This volume.

Ross, M.D. and H.-R. Gregorius 1983. Outcrossing and sex function in hermaphrodites: A resource-allocation model. American Naturalist 121, 204-222

Weir, B.S. 1970. Equilibria under inbreeding and selection. Genetics 65, 371-378

Ziehe, M. 1982. Zygotic genotypic frequencies under selection on female or male gamete production in partially self-fertilizing plant populations. Heredity 49, 271-290

Ziehe, M. 1983. Genotypic frequencies of the offspring generation under selection on female or male gamete production in partially self-fertilizing plant populations. Göttingen Research Notes in Forest Genetics No. 5, Göttingen

Ziehe, M. 1984. Existence of polymorphic genotypic equilibria under asymmetry of homozygotic disadvantage and a deficiency of heterozygotes. Heredity 52, 337-346

Ziehe, M. and H.H. Hattemer 1984. Neuere Erkenntnisse über Asymmetrie-Effekte in der sexuellen Reproduktion von Waldbäumen. Allgemeine Forst- u. Jagdztg., in print.

MATING SYSTEM DYNAMICS IN A SCOTS PINE SEED ORCHARD

W.M. Cheliak

ABSTRACT

Isozymes from viable embryos were used to study the distribution of the effectively incorporated pollen pool in a Scots pine seed orchard. Although selfing was not a significant component of the mating system in this orchard, several forms of non-random mating were observed. These included possible non-random mating patterns marked by one locus, heterogeneous pollen pool distributions, and possible asynchronous sexual phenology schedules among the various clones. Implications of these distortions on the genetic structure and effective population size of the filial generation are discussed.

INTRODUCTION

Clonal forestry, or the use of asexually derived propagules in artificial regeneration programs, is expected to play an increasing role in overall forest production (Libby and Rauter 1984, Verma and Einspahr 1983). However, before significant numbers of these propagules can be used in an operational program, many problems, both genetic and non-genetic, have to be solved (Bonga and Durzan 1982). Thus, within the near future, and certainly for the immediate needs of many tree improvement programs, propagules derived sexually will be predominant for artificial regeneration activities. For most tree improvement programs, these propagules will be obtained from seed production areas or, preferably from seed orchards containing progeny tested clones. For the purposes of this paper, however, we will restrict our discussion to seed orchards.

Clones represented in a seed orchard are expected to reflect an underlying genetic improvement for the suite of traits that have been selected. Thus, we can think of these clones as packages of genetic information which we would like to reflect in the progeny. To ensure that they fully represent the genetic structure of the parents panmixia is required. Therefore, random mating without intervening selection is necessary. Additionally, for monoecious species, this necessitates that all clones contribute and incorporate equal proportions of gametes in the male and female gamete pools. Otherwise, sexually asymmetric contributions will occur which, for our purposes, are equivalent to assortative mating systems.

Seed orchards, especially when isolated from populations of the same species, are effectively finite populations. This is because genotypes of all the clones can be obtained, and the total gametic output of these clones can be unambiguously defined. With these data, several important parameters can now be estimated. These include estimates of levels of contamination from outside sources (Nagasaka and Szmidt 1985, Smith and Adams 1983) and, perhaps more importantly, the assignment of paternity (Elstrand 1984). In addition, arrays of open-pollinated progeny sampled from single trees provides an opportunity to estimate mating system parameters (Brown and Allard 1970).

In this paper, we will concentrate upon the assignment of paternity, and, subsequently, upon the dynamics of the effectively incorporated pollen pool in a Scots pine seed orchard. In addition, we are interested in assessing the homogeneity of the pollen pool in this seed orchard.

MATERIALS AND METHODS

Seed collection, extraction, and storage

Cones were collected in the fall of 1982, and seeds extracted using routine procedures developed at the Petawawa National Forestry Institute. After extraction, seeds were stored under standard moisture and temperature conditions in the seed bank maintained at the Institute. Most cone collections were bulked by ramet, but kept separate by maternal parent clone. However, seed collections from two clones, 909 and 844, were kept separate by ramet as well. Information derived from the bulked collections will be used to compare estimates of variability in the parental and filial generations. To prepare the bulked sample, a number of seeds were analyzed from each clone, weighted by the number of ramets which contributed to the collection. Information from the single ramet collections of the two clones will be used to study dynamics and distribution of the effectively incorporated pollen pool. To study these dynamics, we have analyzed approximately 20 viable embryos from each of the ramets of two clones in the orchard. The maternal haploid gamete, and corresponding diploid embryo genotype, was recorded to enable reconstruction of the multilocus paternal gamete.

Electrophoresis

In March, 1984, seeds were withdrawn from the seed bank for the current study. Seeds were germinated with alternating temperatures of 20°C (16 h) and 30°C (8 h), and light during the high temperature regime. Megagametophytes and corresponding embryos, with 2-5 mm of radicle extension, were placed singly in 0.5 mL conical polystyrene cups (Elkay Products, Inc., Shrewsbury, MA). Female gametophytes were ground at 4°C in 70 μL of an extraction buffer described by Yeh and O'Malley (1980), and embryos were gound at 4°C in the above buffer with 15% (v/v) of the extraction buffer described by Cheliak and Pitel (1984). All tissues were ground with a power-driven teflon grinding head. Protein from the crude homogenates was absorbed on 11 x 1.5 mm wicks cut from Whatman 3MM grade filter paper.

Starch gels, 12.5% (w/v), were prepared from a 1:1 mixture of Electrostarch (Electrostarch Co., Madison, WI, USA: lot 392) and Connaught starch (Connaught Laboratories, Willowdale, Ontario, Canada). Elecrophoresis proceeded until the tracking dye had migrated 6-7 cm. Gels were then removed, sliced horizontally into 1 mm thick sections and stained for the enzymes described in Cheliak and Pitel (1984). Two buffer systems, histidine-citrate (H) and lithium hydroxide-borate (B), described by Cheliak and Pitel (1984) were used to resolve the loci observed to be variable among the clones for this study.

History of the orchard

Clones for this orchard were originally selected for a test of their resistance, or susceptibility, to attack by white pine weevil (*Pissodes strobi* L.) in 1964. Simultaneously, the ortets were also selected for their form as potential Christmas trees. Thus, this orchard has been used for seed production of genetcally improved Christmas trees. The first commercial crop of seed was harvested from twelve clones in the fall of 1982, and during the winter of 1982 the orchard was thinned to maintain only the thirteen most desirable clones. For those clones that were removed from the orchard during the thinning, genotypes were determined from ramets maintained in a clone bank. Thus, we were able to reconstruct the entire genetic composition of the orchard at time of pollination in 1981.

Scots pine is an introduced species to North America. Thus, no natural stands are present to contribute genes to the pollen pool of this orchard. However, within this experimental area, there are two sources of Scots pine within 500 m of the orchard. Although both are genetically related to the orchard clones, there are additional families and clones maintained in these gene banks. Therefore, there is a possibility of contamination. However, because of the similarity of the gene resources in these areas, the probability of identifying contamination will be low. In addition to this obvious source of contamination, there were isolated ramets of several clones which were obviously not members of the clone. It is not clear where this error had occurred. However, if these ramets successfully contribute pollen to the next generation without our knowledge of their identity, it could appear as contamination.

Phenology rating

The ideal of panmixia can be obtained only if the individuals participating in the mating event are synchronized in their sexual phenology. To estimate the degree of synchronization, we used a qualitative index to rate the development of the male and female flowers for each of the clones in the study. These data represent the mean value from two years of observation. No changes in ranking were observed between years.

RESULTS

Levels of variation and genotypic structure

Six loci were observed to be variable among the 12 Scots pine clones surveyed in this orchard. Two alleles were observed at each of five loci, and three alleles at Aat-2 (Table 1). No additional alleles or unique genotypes were observed among the other clones that were reminant in the orchard at the time of pollination of this crop. Levels of expected heterozygosity ranged from a low of 0.08 for Aat-1 to 0.49 for Aat-3, with means of 0.35 and 0.36 in the parental and filial populations, respectively. No additional alleles were detected in the filial generation, indicating that, at the level of unique alleles, no contamination was observed. No significant differences were observed between estimates of the allele frequency in the mature and filial generations. The maximum difference observed was only about 4%. No obvious changes occurred reflecting a consistent increase or decrease in the frequency of common or rare alleles. The same picture, of no systematic changes, emerges when one compares levels of expected heterozygosity in the two generations.

The observed genotypic distribution of the parental and filial generation conformed to that expected under Hardy-Weinburg equilibrium for all loci except 6Pg-1. A significant excess of heterozygotes was observed in both generations for this locus. Since these are clones assembled for breeding purposes, the departure from Hardy-Weinburg equilibrium expectations for one locus in the parental generation is not unexpected. However, one generation of random mating among clones in the parental generation would bring this locus back into equilibrium. Therefore, the significant excess of heterozygotes observed in the filial generation for 6Pg-1 can be taken as evidence for some type of non-random mating patterns. Whether these act on the locus itself, or on a linkage block with which it is in disequilibrium, is not clear. Further, we do not know whether this apparent effect of "non-random" mating takes place during the gametic stage (disassortative mating patterns), or the zygotic stage (selection by lethal allelism). Results for the other five loci suggest that they could be considered approximately neutral with respect to the phenotypic selection criteria used for this population.

Pollen pool distribution

A simple hypothesis of random mating with equal contribution, incorporation, and distribution of male gametes by the orchard clones, would predict the effectively incorporated pollen pool to be homogeneous over the array of plants in the seed orchard. Two sources of evidence argue against this simple model of pollen pool movement and homogeneity in this seed orchard.

Firstly, Aat-2 in the effectively incorporated pollen pool for clones, 844 and 909, indicated significantly heterogeneous distributions among ramets in the orchard (Table 2). As well, Aat-3 indicated significant heterogeneity among the ramets of clone 844, and marginal significance (10%) among the ramets of clone 909.

Table 1. Summary of genetic characteristics observed for the parental and filial generations of a Scots pine seed orchard for six variable loci.

Locus	Allele/Parameter	Maternal	Filial	Paternal
Aat-1	1	.96	.96	.97
	2	.04	.04	.03
	h	.08	.08	.06
	F	-.06	-.05	--
	var (F)	.01	.01	--
	N	12	191	191
Aat-2	1	.75	.77	.56
	2	.21	.23	.44
	3	.04	--	--
	h	.39	.35	.49
	F	-.27	-.05	--
	var (F)	.16	.01	--
	N	12	185	185
Aat-3	1	.42	.44	.39
	2	.58	.56	.61
	h	.49	.49	.48
	F	-.03	-.06	--
	var (F)	.09	.01	--
	N	12	179	179
Gdh	1	.79	.83	.84
	2	.21	.17	.16
	h	.33	.29	.27
	F	.24	-.06	--
	var (F)	.04	.01	--
	N	12	135	135
6Pg-1	1	.66	.64	.72
	2	.34	.36	.28
	h	.44	.46	.40
	F	-.50*	-.31*	--
	var (F)	.26	.05	--
	N	12	152	152

Locus	Allele/Parameter	Maternal	Filial	Paternal
6Pg-2	1	.66	.64	.66
	2	.34	.36	.34
	h	.44	.46	.46
	F	.25	.01	--
	var (F)	.04	.01	--
	N	12	108	108

* indicates a significant departure a min P[0.05]

Table 2. Summary of log-linear G tests for the homogeneity of allele frequency distribution among ramets of two clones in a Scots pine seed orchard.

	Clone Number			
	844		909	
Locus	G	df	G	df
Aat-2	35.27*	24	40.40*	18
Aat-3	33.34*	12	14.51	9
Gdh	9.93	12	10.52	9
6Pg-1	-	--	8.14	5
6Pg-2	16.17*	9	6.10	9

* indicates significant departure at [P 0.05]

The distribution of <u>6Pg-2</u> was significantly heterogeneous among the ramets of clone 844. Thus, 75% of the loci indicated a significantly heterogeneous distribution among the ramets of clone 844, and 20% among the ramets of clone 909.

Secondly, consider the incorporation of unique allele 3 of <u>Aat-2</u> from clone 889. If the pollen pool were homogeneous one would expect that the frequency of this unique allele would be independent of the distance between a marker and source ramet (Figure 1). Although the relationship is far from perfect, one does not observe any unique alleles in ramets which are separated from the source by more than three or four positions in any direction. Alternatively, when we see the rare allele, its frequency tends to be higher in those marker ramets which are closer to the source (e.g. 844 rows 11 and 12 versus rows 3, 5, or 7; Table 3).

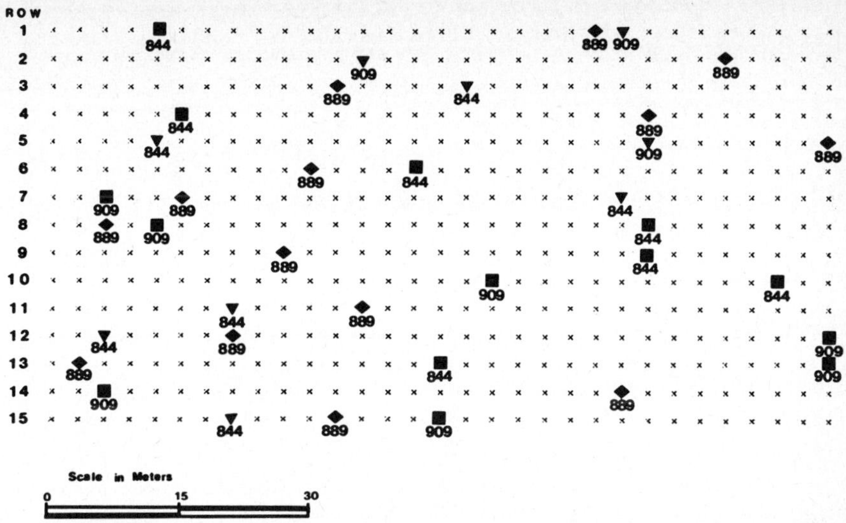

Figure 1. Distribution of unique pollen allele Aat-2 #3 from source clone 889 recovered from viable embryos of ramets from marker clones 844 and 909. (◆)=source ramet 889, (▼)=unique allele observed, (■)=unique allele not observed. An x indicates the location of other ramets in the orchard at the time of original establishment.

Table 3. Allele frequency variation at Aat-2 for ramets of clones 844 and 909 in a Scots pine seed orchard.

		ROW														
Clone	Allele	1	2	3	4	5	6	7	8	9	10	11	12	13	14	15
844	1	1.	-	.65	.80	.85	.70	.68	.85	.40	.74	.75	.70	.72	-	.90
	2	-	-	.30	.20	.10	.30	.26	.15	.60	.26	.15	.20	.28	-	.05
	3	-	-	.05	-	.05	-	.06	-	-	-	.10	.10	-	-	.05
909	1	.67	.42	-	-	.78	-	.90	.20	-	.80	-	.78	.50	.79	.95
	2	.30	.50	-	-	.11	-	.10	.80	-	.20	-	.22	.50	.21	.05
	3	.03	.08	-	-	.11	-	-	-	-	-	-	-	-	-	-

Taken together, these results support the hypothesis of restricted effective transmission distances of the pollen pool (Müller 1977, Müller-Starck 1979, Shen *et al.* 1981). In this orchard, the minimum effective transmission distance appears to be 7.4±0.8 m considering unique alleles and unique multilocus gametes, with the criterion of distance to nearest source. Implicit in this result, and supported by the heterogeneity statistics, is that the genetic composition of the pollen pool changes significantly over the orchard. Thus, in this orchard, the concept a neighborhood, with a maximum effective size of 21 individuals, would appear to describe the biologically effective pollen pool from which an individual ramet would likely derive its pollen.

Multilocus structure and paternity identity

Genotypes of the 12 maternal parent clones were inferred from analysis of haploid megagametophytes and determined directly by somatatyping. Each of these clones can be uniquely identified on the basis of their multilocus diploid genotype (Table 4). This result demonstrates the power of electrophoresis as an aid to genetic identity when working on populations with high levels of polymorphism. With the allelic composition of these six variable loci, a total of 68 different multilocus gametes can be produced. Twenty-four of these 68 haploid gametes are unique, and can be assigned to a single clone. Thus, even though only one clone has a unique allele (Clone 889 for allele 3 at Aat-2), considerable genetic identity is possible by exploiting the rich multilocus structure of this orchard.

Table 4. Diploid genotypes, at six variable enzyme loci, of twelve clones in a Scots pine seed orchard.

	Parent Clone Number											
Locus	418	420	613	844	845	889	899	900	907	909	1134	1246
Aat-1	11	11	11	11	11	11	12	11	11	11	11	11
Aat-2	12	12	11	11	11	13	12	11	12	11	12	11
Aat-3	12	12	22	11	11	12	12	12	22	12	22	22
Gdh	12	11	11	11	11	11	11	22	12	12	11	11
6Pg-1	12	11	12	12	11	12	12	12	11	12	11	12
6Pg-2	11	12	11	12	11	11	12	12	11	22	11	22

With the maternal haploid contribution and the corresponding diploid embryo genotype, the multilocus paternal gamete is ascertained by inspection. Once the paternal contribution has been determined, this multilocus genotype is compared with the gametic array that could potentially be produced by the known parents. Paternity is then assigned to those individuals that could have produced the multilocus gamete. When more than one individual is implicated as a possible male, it is possible to assign the most likely parent by constructing indices which estimate the likelihood of paternity, such as:

L={Pr[Producing gamete]/[Frequency of gamete in the population]}

Many other refinements are possible, and are dealt with in detail by Lee *et al.* (1980), and Heise *et al.* (1983), for example.

In this paper, we shall concentrate upon those cases where unambiguous paternity can be assigned from the multilocus gametes. Of the total clonal composition at the time of pollination, three clones, 418, 889, and 899, produce unique gametes. Clones 418 and 889 each produce four unique gametes, and 899 produces 16 unique gametes. If these clones contribute equal numbers of gametes to the pollen pool, we could expect that the unique gametes would be recovered in a 1:2:2 ratio from the viable progeny of our marker clones. When we pool the data for the ramets of both clones, and test this expectation, we find no departure from the expected distribution (G=1.01, 2 df). Thus, all three clones appear to be contributing gametes at the expected proportions, at least by the evidence obtained from these two marker clones.

We have not attempted to assess net fertility of orchard clones by analysis of bulked seed samples (Müller-Starck 1982, Müller-Starck *et al.* 1983). This is because seed samples were obtained only from twelve maternal clones, and are therefore not representative of the total genetic structure of the orchard. Therefore, it is not clear from these results whether the sexually asymmetric contributions observed by others (Müller-Starck 1982, Müller-Starck *et al.* 1983) are a factor in this orchard. However, it is likely that in addition to the biological factors identified as contributing to the sexual asymmetries, asynchronous sexual phenology schedules among the clones also contributes to these effects (Figure 2). In subsequent analyses of the mating system dynamics of this orchard, we hope to be able to separate the confounded effects of asynchronous sexual phenology schedules and sexual asymmetries.

SUMMARY AND CONCLUSIONS

Preliminary results indicate that selfing is not a significant problem among, or within, the ramets of clones in this orchard. However, other components of random mating (i.e., panmixia) which require equal likelihoods of mating among all individuals are most likely violated in this orchard. Several deviations of this type were observed. These involved some type of non-random mating marked by one locus (*6Pg-1*), as well as asynchronous sexual phenology schedules among the orchard clones, heterogeneous pollen pool distributions, and probable restricted effective transmission distance of pollen from any source. Results indicate that the average minimum transmission distance in this orchard is approximately 7.4 m.

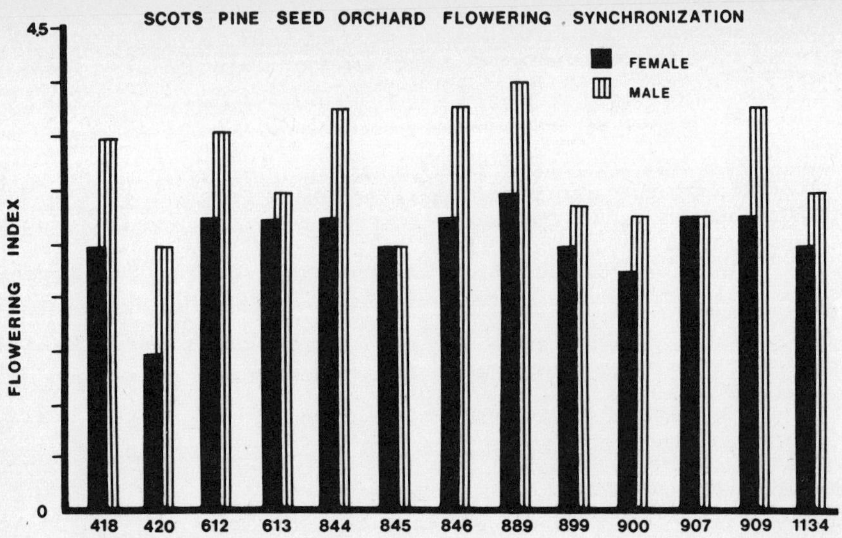

Figure 2. Average index for male and female sexual phenology rating at maximum female receptivity of the marker clones in a Scots pine seed orchard. Numerical rating is as follows: When squeezed, the male flowers were; 1=wet, 2=damp, 3=dry, or 4=the pollen was flying. An average value of 3 is usually considered sufficient for significant pollen dispersal. For females, the rating was; 1=breaking bud, 2=opening, but not receptive, 3=open, receptive, 4=closed.

Asynchronous phenology schedules, confounded with asymmetrical fertility, can have serious impacts on the effective population size of the zygotes derived from seed orchards. Maximization of effective population size will occur when the number of genetically effective males equals the number of genetically effective females. As well, the genetic structure and relationships among cohorts of the filial generation will change as these deviations become more pronounced. Optimization of genetic diversity for regeneration purposes will occur with orchard designs which maximize the neighborhood diversity around ramets with synchronous phenology schedules.

ACKNOWLEDGMENTS

The author would like to acknowledge the excellent assistance of Mssrs. P. Copis, E. Gilchrist and J. Pitel in completing this study. The critical reading and excellent suggestions by an anonymous reviewer were especially valuable for the final form of this paper.

LITERATURE CITED

Bonga, J.M. and D.J. Durzan (eds). 1982. Tissue culture in forestry. Martinus Nijhoff/Dr. W. Junk Publishers. The Hague.

Brown, A.H.D. and R.W. Allard. 1970. Estimation of the mating system in open-pollinated maize populations using isozyme polymorphisms. Genetics 66:133-145.

Cheliak, W.M. and J.A. Pitel. 1984. Genetic control of allozyme variants in mature tissues of white spruce trees. J. Hered. 75:34-40.

Ellstrand, N.C. 1984. Multiple paternity within the fruits of the wild radish, Raphanus sativus. Amer. Natur. 123:819-828.

Heise, E.R., C. Keever and M.R. McMahan. 1983. A critical analysis of paternity determination using HLA and five erythrocyte antigen systems. Am. J. Forens. Med. and Pathol. 4:15-23.

Lee, C.C., L.K. Lebeck and C. Wong. 1980. Estimating paternity index from HLA-typing results. Am. J. Clin. Pathol. 74:218-223.

Libby, W.J. and R.M. Rauter. 1984. Advantages of clonal forestry. For Chron. 60:145-149.

Müller, G. 1977. Cross-fertilization in a conifer stand inferred from enzyme gene-markers in seeds. Silvae Genet. 26:223-226.

Müller-Starck, G. 1979. Estimates of self and cross-fertilization in a Scots pine seed orchard. Proc. Conf. Biochem. Gent. of Forest Trees, Umeå, Sweden 1978.

Müller-Starck, G. 1982. Sexually asymmetric fertility selection and partial self-fertilization. 2. Clonal gamete contributions to the offspring of a Scots pine seed orchard. Silvae Fennica 16:99-106.

Müller-Starck, G., M. Ziehe and H.H. Hattemer. 1983. Reproductive systems in conifer seed orchards. 2. Reproductive selection monitored at an LAP gene locus in Pinus sylvestris L. Theor. Appl. Genet. 65:309-316.

Nagasaka, K. and A.E. Szmidt. 1984. Multilocus analysis of external pollen contamination of a Scots pine (Pinus sylvestris L.) seed orchard. - Submitted this volume.

Shen, H.H., D. Rudin, and D. Lindgren. 1981. Study of the pollination pattern in a Scots pine seed orchard by means of isozyme analysis. Silvae Genet. 30:7-15.

Smith, D.B. and W.T. Adams. 1983. Measuring pollen contamination in clonal seed orchards with the aid of genetic markers. Pp. 69-77 in Proc. 17th Southern Forest Tree Impr. Conf. Univ. of Georgia, Athens.

Verma, D.C. and D.W. Einspahr. 1983. Conifer tissue culture and how it may impact the pulp and paper industry. Tappi 66:25-27.

Yeh, F.C.H. and D. O'Malley. 1980. Enzyme variations in natural populations of Douglas-fir, Pseudotsuga menziesii (Mirb.) Franco, from British Columbia. 1. Genetic variation patterns in coastal populations. Silvae Genet. 29:83-92.

REPRODUCTIVE SUCCESS OF GENOTYPES OF PINUS SYLVESTRIS L. IN DIFFERENT ENVIRONMENTS

G. Müller-Starck

ABSTRACT

To quantify the reproductive success of individual genotypes in different environments, the realized female and male fitness values were estimated for single clones in each of two spatially and two temporally different environments. Each of the studied clones carries a unique multilocus genotype enabling the determination of its contribution to the offspring. The results are based on 1o-locus genotyping of the parental clones and the four seed orchard offspring populations, represented by 5oo seeds each.

The reproductive success of the clones differ substantially from one another, regardless of whether the female, the male or both fitness components of a clone are concerned. Strong sexual asymmetries are observed which also occur in sexually opposite directions. The results do not support the assumption of genetically controlled reproductive efficiencies. However, in some cases it cannot be ruled out that this is due to environmentally dependent selection with respect to the sex function of single clones. Some consequences of the observed phenomena are outlined briefly.

Key words: Reproduction, realized fitness, sexual asymmetry, multilocus, clone, pine seed orchards

INTRODUCTION

The reproductive success of individual genotypes is measured in terms of realized fitness values (Gregorius and Ross 1981), which are defined by the average number of the successful gametes. A female or male gamete is considered to be successful if it is incorporated in a zygote. The census stage is that of fully developed seeds after open pollination. Thus the progeny generation is described at the embryo stage. The realized fitnesses used here refer to single annual reproductive periods (realized annual fitness values, Müller-Starck and Ziehe 1984).

The reproductive success was studied in clonal seed orchards, i.e., populations in which the parental trees (clones) are represented by several genotypically identical replicates. Thus the information obtained for each clonal genotype is the average value of the sampled replicates. For Scots pine (Pinus sylvestris L.), clonal seed orchards are an important medium for seed production, the seed crops of which are being used to an increasing extent in place of seeds from natural forest stands.

In order to monitor environmental impacts on the reproductive success of clonal genotypes, an experimental situation consisting of different spatial and temporal environments was chosen. As a first step, the reproductive success of representatives of single clones were studied in two geographically separated seed orchards and two flowering periods. For each of the four seed crops, the multilocus genotypes among the embryos and corresponding endosperms were compared with the genotypes of the parental clones, so that their female and male contributions to the offspring could be identified.

It is the aim of this study to proceed in the characterization of reproduction in pine breeding populations (e.g. Hadders and Koski 1975, Sweet 1975, Jonsson et al. 1976, Rudin and Lindgren 1977, Adams and Joly 198o, Moran et al. 198o, Chung 1981, Shen et al. 1981, Friedman and Adams 1982, Squillace and Goddard 1982, Schmidtling 1983, Ross 1984, Müller-Starck 1979, 1982 a,b). In particular it is intended to make a contribution in answering the general question as to the extent to which environmental changes can affect the reproductive success of genotypes in populations.

MATERIAL AND METHODS

Seed orchard populations

The two seed orchards under study were established using the same clones in 1958 by the Bayerische Landesanstalt für forstliche Saat- und Pflanzenzucht, Teisendorf. They are situated in the area of Ebrach, Bavaria, in the forest districts Winkelhof/Schafknock (=I) and Ebrach 1/Herrgottschlag (=II) at an altitude of 38o m (I) and 33o m (II) above sea level. The distance between I and II is 5 km. The initial number of clones was 45, with a total of 8oo representatives (population size) in I and 616 in II. Due to snow breakage and thinnings, the clone number was reduced to 42 in I and 39 in II, with population sizes 562 in I and 274 in II, and the number of clones in common was reduced from 45 to 38. These values remained constant over the studied flowering periods. In spite of the initial block design, the representation of the clones in the total population is very heterogeneous in both seed orchards and varies between 0.4 and 11.8%. The seed orchards are located in an area predominantly covered with beech stands and thus are not directly exposed to external pine pollen influx.

Seed samples

A set of four seed samples (from two seed orchards over two flowering periods) was studied, each comprising 5oo seeds. The probes originate from routine all-orchard harvests collected in both orchards in 1979 and 1983 and thus refer to the open pollination during the flowering periods 1978 and 1982. The seeds were supplied by the Bayerische Landesanstalt für forstliche Saat- und Pflanzenzucht. The samples were randomly drawn from the bulked seed orchard crops.

Genotyping

Multilocus-genotypes were identified by applying horizontal starch gel electrophoresis and utilizing the enzyme systems leucine aminopeptidase, glutamate oxalacetate transaminase, malate dehydrogenase, shikimate dehydrogenase (LAP, EC 3.4.11.1; GOT, 2.6.1.1;MDH, 1.1.1.37; SKDH, 1.1.1.25); the genetic control was verified beforehand by analyzing seeds from controlled crossings (Müller-Starck 1982, a,b and unpublished). The genotypes were scored simultaneously at 1o gene loci: LAP-B, GOT-A,B,C, MDH-A,B,C,D, SKDH-A.B. Like the other loci, MDH-B and MDH-C segregate independently, but form a heterodimer in haploid tissue (see also O'Malley et al. 1979 and El-Kassaby 1981); in diploid tissues different heterodimers are obtained as simple combinations, whereas the allelic variation within MDH-B and MDH-C results in hybrid bands due to the dimeric state of MDH in Scots pine.

The genotypes of the parental clones were identified by endosperm analyses using up to five individuals per clone and 1o seeds per individual. In some cases, trees in the orchard labeled as replicates of one and the same clone were proven to have different genotypes, so that in order to avoid misclassifications, the 1o-locus genotype of each individual tree in both orchards had to be determined. Because the phenotypic expression of the applied enzyme genotypes is tissue-unspecific, the genotyping could be done on buds or needles (for method see Müller-Starck 1982 a, with additional freeze drying).

The genotypes of the seed orchard offspring were identified by means of analysis of endosperm and corresponding embryo of each individual seed, so that the female and male contribution could be determined as an ordered paair (Müller (-Starck) 1976).

Clonal gametic contributions to be offspring

Clones were selected which carry alleles or multilocus genotypes which are unique in both seed orchards and thus produce identifiable gametic contributions to the all-orchard offspring. By comparing the clonal genotypes with those of the seeds, the respective clones then can be identified as female or male parent or both. The program GSED ("Genetic Structures from Electrophoresis Data", Gillet unpublished) was applied and additional search runs programmed (Radler unpublished). If a clone is homozygous for a unique allelic marker, its entire female and male contribution can be determined. If the allelic marker is in the heterozygous state, only a portion is identifiable, the size of which depends on the minimum number of loci necessary to identify the unique status of the respective parental clone (e.g. 0.5 in case of one locus, 0.25 for two loci, etc.).

The following assumptions were made: Firstly, that the applied gene loci are unlinked and do not show deviations from the Mendelian 1 : 1 segregation within zygote contributions. Observations in full sib families support these assumptions (Müller-Starck 1979, 1982 a and unpublished). Secondly, that the genotypic survival rate between the early zygote stage and that of fully developed seeds is assumed to be the same for all zygotes. Finally, that the influx of external Scots pine pollen is negligible or results in genotypes which can be distinguished from those originating from the orchard.

Fitness values

Estimates of the realized annual fitness values were based on the calculated female and male contributions of the clones to the all-orchard offspring. Because the total amount of zygotic production is unknown, only fitnesses relative to the population mean are used (Müller-Starck and Ziehe 1984). For each clone this mean value is equal to the respective clonal contribution under panmictic conditions. Due to the different number of replicates per clone in the seed orchards, the expected panmixia values were calculated separately for each clone. The given female and male contributions then were divided by the respective expected panmictic value, and the obtained fitnesses were plotted for the given temporal and spatial environments.

RESULTS AND DISCUSSION

Clonal gametic contributions

Of the 37 clones represented in both seed orchards, at least 16 clones carry a unique 10-locus genotype and thus produce identifiable gametes. The unique genotypic status of these clones is based on not more than three heterozygous loci with one marker allele apiece: clones No. 3, 17, 18, 26, 39, 45 one locus; clones 6, 7, 9, 22, 25, 27, 40, 43 a combination of two loci; clones 11 and 38 a combination of three loci (the complete array of multilocus genotypes is given in the appendix). Consequently, the detected number of seeds in the all-orchard offspring which contain the female or male contribution of a particular clone had to be multiplied by the factors 2, 4 and 8, respectively (clones belonging to the third category are not included in the estimation of fitness values (see Fig. 1-3)). The resulting estimated numbers of seeds for each seed orchard and flowering period are presented in Table 1 along with the mean gametic contributions and the number of seeds which can be expected under the panmixia hypothesis ($C_i/N \times 500$ with C_i = number of representatives per clone i, N = population size, 500 = number of seeds per sample).

In Table 1 the results concerning all identifiable clones are included, independently of whether or not the sample sizes of 500 are sufficient to reveal a significant gametic representation. Some of the clones have too few replicates (see expected values under panmixia of the clones 11, 22 (I), 26 (I); some are underrepresented as compared to the panmixia values (e.g. clones 9, 22 (II), 40 (II)). Not taking into account the first category, some obvious phenomena can be observed:

Firstly, the mean gametic contributions deviate in several cases significantly (2 x 2 homogeneity test) from the corresponding values under panmixia: Overrepresentation is obtained, for instance, in the flowering period 1982 for the clones 7, 18, 25 in II; among those underrepresented are 9 (I) and 22 (II) in both periods, 17 (II) in 1982, 18 in 1978.

Table 1. Estimated number of seeds with the female or male gametic contribution of identifiable parental clones in four all-orchard crops (two seed orchards in two flowering periods; sample size 4 x 500 seeds). The given values result from multiplication of the number of identified seeds by the factor 2 or 4 (see article) and only in the case of the clones No. 11 and 38 by the factor 8. For comparison, the corresponding values expected under panmixia are given for each seed orchard and clone.

DESIGNATION OF IDENTIFIABLE CLONES	CLONAL GAMETIC CONTRIBUTIONS IN ORCHARD CROPS FROM 2 FLOWERING PERIODS													
	SEED ORCHARD I							SEED ORCHARD II						
	♀,♂ UNDER PANMIXIA	♀ CONTRIBUT. 1978 → 82		♂ CONTRIBUT. 1978 → 82		MEAN CONTRIBUT. 1978 → 82		♀,♂ UNDER PANMIXIA	♀ CONTRIBUT. 1978 → 82		♂ CONTRIBUT. 1978 → 82		MEAN CONTRIBUT. 1978 → 82	
3	8.9	6	8	18	4	12	6	9.1	4	4	6	12	5	8
6	8.9	4	4	12	16	8	10	14.6	0	20	12	12	6	16
7	15.2	4	16	32	24	18	20	9.1	8	48	8	20	8	34
9	23.2	0	8	4	4	2	6	7.3	0	0	0	0	0	0
11	1.8	8	0	8	0	8	0	1.8	0	0	0	0	0	0
17	41.9	72	34	4	10	38	22	27.4	36	4	2	0	19	2
18	13.4	2	14	0	4	1	9	21.9	8	106	2	8	5	57
22	1.8	0	0	0	0	0	0	18.2	4	0	0	0	2	0
25	3.6	12	4	4	0	8	2	3.6	8	20	0	8	4	14
26	1.8	0	0	0	0	0	0	23.7	24	24	6	12	15	18
27	12.5	32	4	8	12	20	8	7.3	12	8	4	4	8	6
38	7.1	0	40	8	8	4	24	7.3	8	0	8	8	8	4
39	10.7	14	22	0	2	7	12	9.1	8	0	2	0	5	0
40	7.1	8	0	0	0	4	0	10.9	0	12	0	4	0	8
43	8.9	4	8	0	0	2	4	3.6	0	16	0	0	0	8
45	4.5	6	0	2	2	4	1	10.9	8	0	4	0	6	0

Secondly, the frequencies of the female and the male contribution of one and the same clone can deviate highly significantly from each other, mostly in favour of the females: for example clone 17 with ratio females: males of 72:4 (I) and 36:2 (II) in 1978, clone 18 (II) with 106:8 in 1982, clone 27 (I) with 32:8 in 1978; in favour of the male contributions were e.g. clone 6 (II) with 0:12 in 1978 and clone 7 (I) with 4:32 in 1978. These asymmetries in the female and male sex function confirm the results of earlier studies (Ross 1984, Müller-Starck 1982a, Müller-Starck et al. 1983).

Thirdly, within each seed orchard the clonal gametic contribution can deviate substantially between the two flowering periods; with respect to the female contribution e.g. clone 7 (II) with 8 in 1978 versus 48 in 1982, clone 17 (II) with 36 versus 4, clone 18 (II) with 8 versus 106, clone 27 (I) with 32 versus 4. The differences in the male contributions are less pronounced.

Finally, deviations are evident between the two seed orchards: For instance, in 1982 the mean gametic contribution of clones 38 and 39 is overrepresented in I as compared to II and of clones 7, 18, 25 vice versa. The obtained deviations can be better assessed by means of the graphically illustrated fitnesses in the subsequent sections.

Reproductive success in different temporal environments

The reproductive success which is expressed by the estimated relative female and male clonal fitness values are given separately for each seed orchard in Figures 1 and 2. These graphs comprehensively illustrate some reproductive characteristics (Müller-Starck and Ziehe 1984). The given fitnesses were calculated by dividing the estimated female and the male contribution of a particular clone as presented in Table 1 by the respective value under panmixia. A resulting quotient of 1 is equal to the average female or male fitnesses $\overline{W}^{♀}, \overline{W}^{♂}$. Fitness components which meet this condition should be located on the respective subsidiary lines in the applied graphical system. The mean population fitness, which is defined here as the sum of the female and the male fitness, is represented by the straight line connecting the points $2\overline{W}^{♀}$ and $2\overline{W}^{♂}$. The intersection of the two subsidiary lines at $(\overline{W}^{♀}, \overline{W}^{♂})$ is equal to the expected mean female and male fitness under panmixia. The diagonal line from zero through this intersection marks symmetry in the female and male reproductive efficiency. The changes in fitness components from the first flowering period to the second are indicated by arrows.

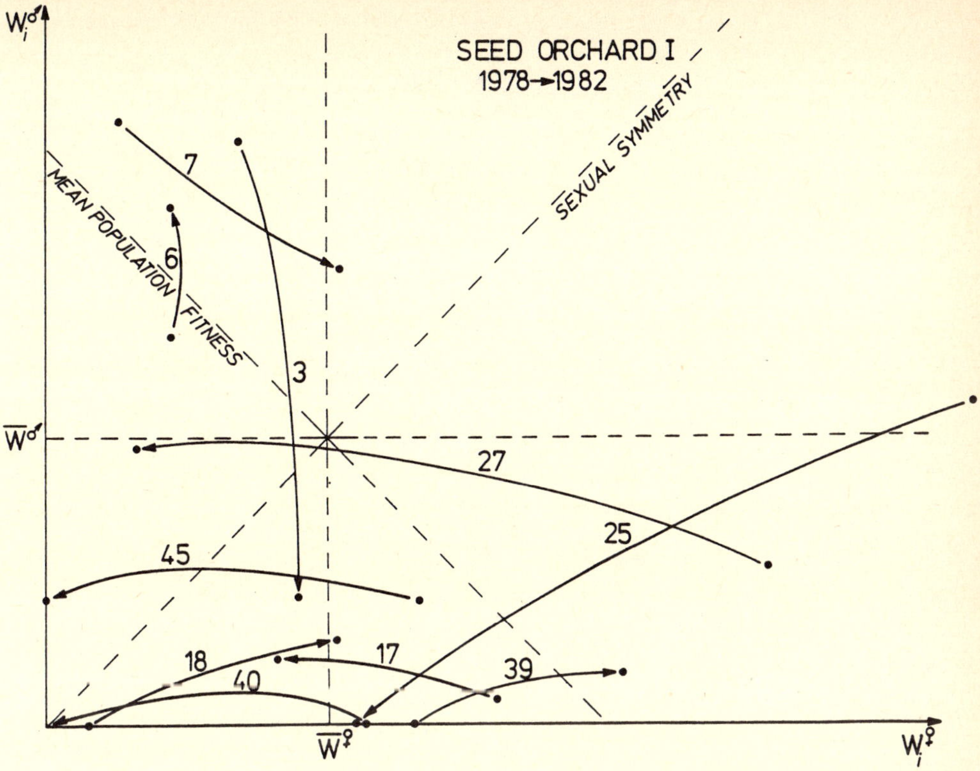

Figure 1: Relative female and male fitnesses of clones i (W_i♀, W_i♂) in two flowering periods (1978→1982) in seed orchard I (Schafknock)

In order to minimize possible biases, in both figures the fitness values of the clones 9, 11, 22, 26, 38 and 43 are not included, since gametic contributions of these clones are too rare and/or the estimates of their frequencies are based on a multiplication factor greater than 4 (i.e., clonal genotypes which obtain their unique genotypic status because of more than two heterozygous marker loci are excluded.

The four phenomena outlined above are adequately demonstrated by the fitness components of the remaining ten clones in Figures 1 and 2. Considerable superiority of the fitness components as compared to the population mean is evident in the flowering period 1978 or 1982 for the clones 7 and 18 in the seed orchard II, 27 in I and 25 in both seed orchards. In the case of the clones 7 and 25 the mean population fitness is exceeded by more than 250% and for clone 18 by more than 150%. In-

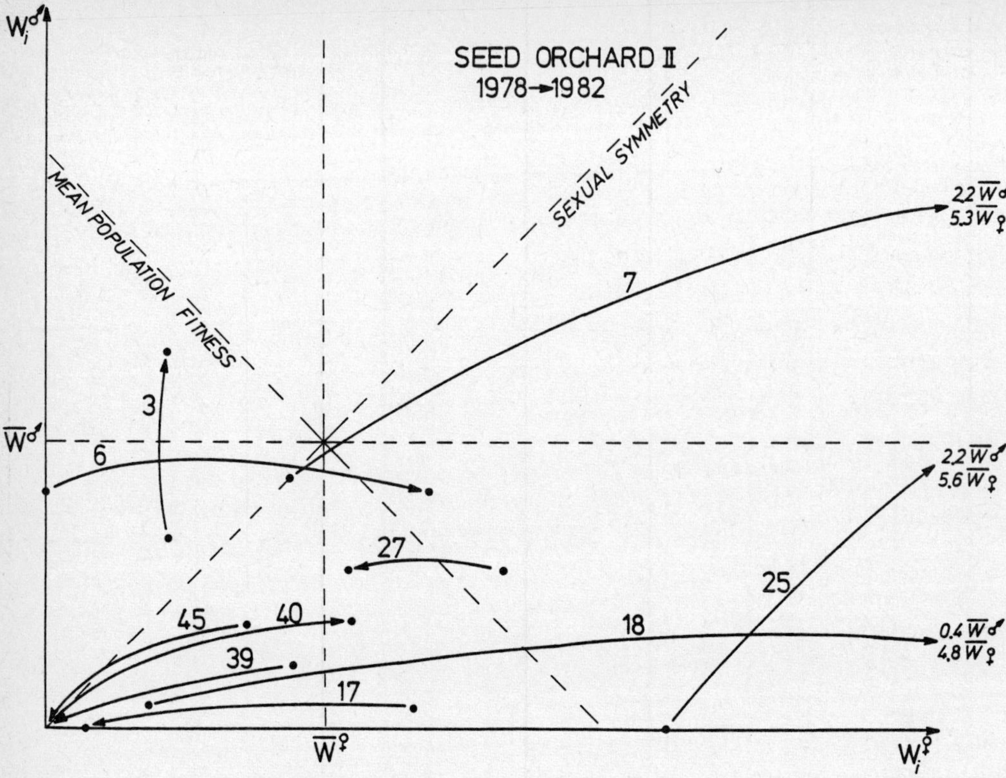

Figure 2: Relative female and male fitnesses of clones i ($W_i^{♀}$, $W_i^{♂}$) in two flowering periods (1978→1982) in seed orchard II (Herrgottschlag)

feriority as compared to the population mean is indicated particularly for the clones 18 (I), 39 (II), 40(I,II) and 45 (II) in both periods but for 6 (II), 17 (I,II), 18 (I,II) and 45 (I) in only one period.

Major emphasis should be put on the deviations between the two flowering periods with respect to the female and the male clonal reproductive success. The obvious phenomenon of asymmetry in the fitness components of most of the clones is, in general, not maintained in the second period. Some clones remain predominantly female or male, such as clones 6, 17, 25, 39 in I and 18, 25, 27 in II. The fitnesses of other clones vary in sexually opposite directions, i.e. clones 3 (I), 6 (II), 27 (I), 45 (I). The occurrence of clonal reproductive success in sexually reversed proportions raises the question as to the extent to which the determination of the sex function of these clones still can be assumed to be under genetic control.

<u>Reproductive success in different spatial and temporal environments</u>

The female and the male fitness values of the clones which are obtained in the same flowering period in the two seed orchards are contrasted in Figure 3. The mean population fitness is marked by two parallel lines. Clonal fitnesses which are identical in both seed orchards must result in a symmetrical figure. In the same temporal but different spatial environment the following tendencies are observable:

In 1978 fitness characteristics of several clones are congruent in both seed orchards: Clones superior in the female fitness component as compared to the mean population fitness show this trend in both spatial environments (clone 17, 25, 27). This is also valid for some clones in in which the female or male fitness was inferior, e.g. 3, 27, 45. Similarities between the orchards seem to be slightly more pronounced in the female sex function. For the majority of the clones congruities in the clonal fitness components between I and II are restricted to one sex only. In general, fitness components appear to be sexually asymmetrical.

In 1982 clonal fitnesses show larger deviations between the seed orchards. Some clones are considerably overrepresented in their female fitnesses but always in only one of the seed orchards (clones 7, 18, 25 in II and clone 39 in I). Similar trends in both orchards are obtained more with respect to the male clonal fitnesses (clones 7, 18, 27) than the female ones (clone 3). Sexual asymmetries are characteristic for the majority of the clones.

Comparing the flowering periods, in 1978 similarities between I and II are more pronounced than in 1982, but with respect to the female and also the male fitness, while in 1982 mainly the male one is favoured. In both periods sexual asymmetries are predominant. Between 1978 und 1982 considerable deviations can be observed: This is obvious, for instance, in the case of a comparison which includes both fitness components of a clone in I and II (e.g. clone 18) or in only one of the seed orchards (e.g. clones 7, 39, 40 in II). Outstanding differences between 1978 and 1982 are indicated with respect to the female and the male reproductive success: The fitness values of a clone in one flowering period appear in sexually reversed directions in the other period. This is valid for the clones 3, 6, 27, 45 in one of the seed orchards.

The results demonstrate that fitness characteristics of single clones generally cannot be assumed to be maintained to a similar extent in varying spatial and temporal environments. There is some indication that

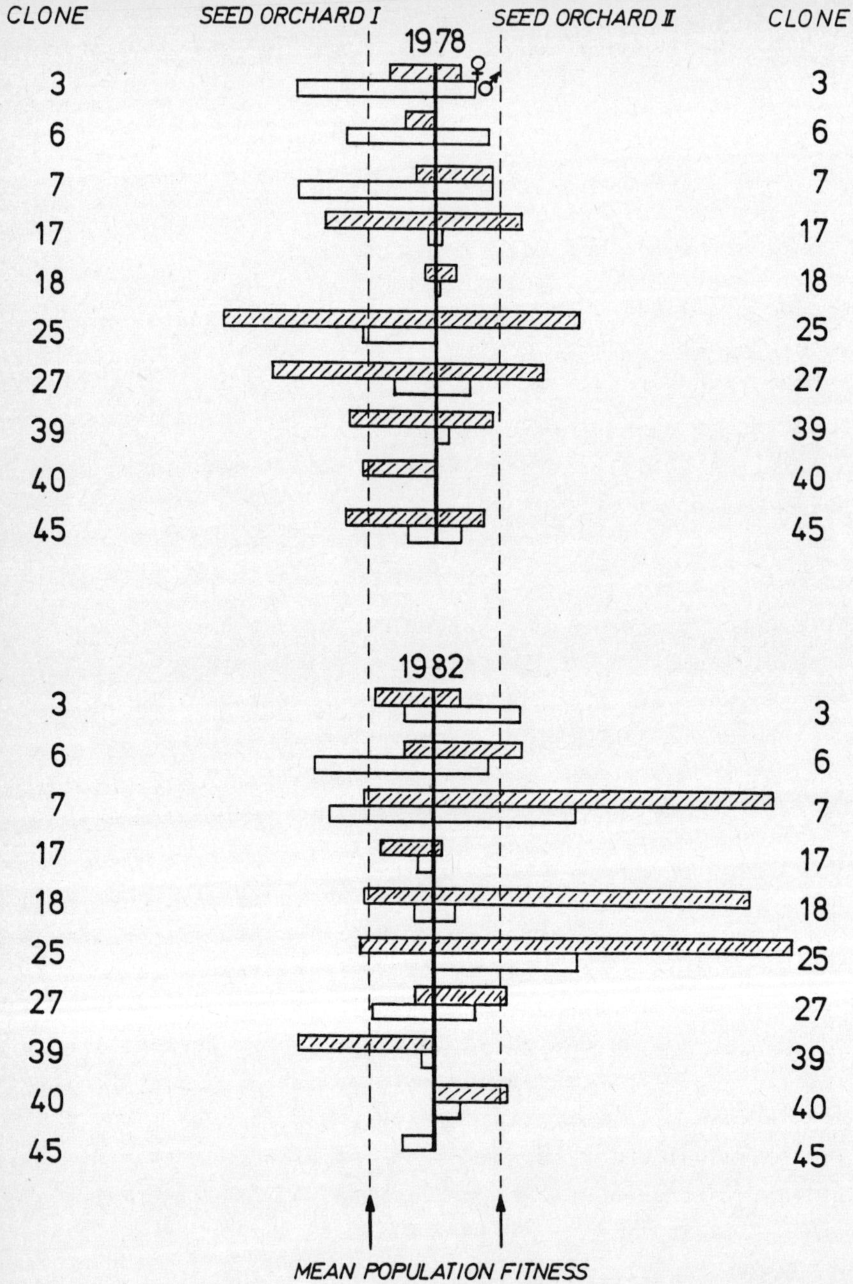

Figure 3: Relative female and male fitnesses of specific clones in two seed orchards (I, II) and two flowering periods (1978, 1982). The female fitness of each clone is indicated by hatched columns, the male fitness by unhatched ones.

the existing similarities in the clonal fitness components are more pronounced in the same temporal but different spatial environment than vice versa.

CONCLUSIONS

Under the given experimental conditions, a remarkable amount of variation in the clonal reproductive efficiencies is obtained between the four environments. The results cannot be unequivocally interpreted as being due to prevailing environmental or ontogenetic impacts. Systematic changes in the fitnesses appear unlikely: In too many cases, the characteristics in the clonal fitnesses are not maintained in the same spatial but different temporal environment and vice versa, regardless of whether the total fitness, the functioning in the female sex or that in the male sex is concerned. Similarities in the clonal fitness components are obtained within one ontogenetic stage (period 1978), but these are not adequately maintained in the other stage. This may be interpreted for the present as evidence against the assumption of a genetically controlled reproductive success of clones. The low fitnesses of some clones might be still concordant with this assumption but not necessarily supportive. The results make it clear that panmictic conditions are far from being fulfilled. If the clones are representative, the non-random mating of genotypes can be assumed to be a prevailing component of the reproductive system in Scots pine seed orchards. As an example, the female fitnesses of the clones deviate from the respective population mean by a factor ranging between zero and 5.5 and the male fitnesses by a factor between zero and 2.2. The respective factor for the mean female and male fitness of the clones range between zero and 3.9. Such large deviations result in a reduction of the reproductively efficient size of the orchard populations and decrease the chances of genotypic similarity between the selected parental clones and their offspring. For the breeding and silvicultural practice, the obtained results suggest that orchard seeds from different flowering periods be mixed as opposed to being handled separately.

In most cases, substantial asymmetries are evident in the clonal female and male reproductive success. For instance, the female fitness component of clone 18 in seed orchard II in 1982 is 4.84 while the male is only 0.37; in the case of clone 7 (I, 1978) the respective proportion ranges from 0.26 to 2.10. It was also demonstrated that changes in sexually opposite directions can occur. The obtained results suggest in some cases the preliminary assumption of environmentally dependent selection with respect to the sex function of single clones. This phenomenon will have to be studied in detail by means of a comparison between the genotypic structures of the parental clones and the four different offspring populations, since, for instance, the occurrence of sexually asymmetrical fertility selection can result in a reduction of the proportion of homozygotes relative to the Hardy-Weinberg proportion and thereby mask inbreeding effects (Ziehe and Gregorius 1981, Ziehe 1982 a,b; examples e.g. Müller-Starck 1982 a).

The given results are a first step in the quantification of the reproductive success of Scots pine clones under varying spatial and temporal environments. The experimental conditions are restricted in such a way that the available set of environments is only 2x2. However, the selected temporal environments represent periods of outstanding flowering in both seed orchards and refer to the same population sizes and conditions. The seed probes originate from full harvests of the entire seed orchard. The modes of harvesting and sampling are the same for all four environments und thus cannot serve to explain phenomena such as the overrepresentation of clones in their female fitnesses in one particular environment (II, 1978). All parental clones fulfil the requirement that both sexes flower, which is an important criterion in the selection of such clones for breeding purposes. To reduce biases in the estimation of fitness values due to small sample sizes, the study of the reproductive success was restricted to 10 of the 16 identifiable clones represented in both seed orchards. The full information potential of the lo-locus genotypes could have been utilized only by applying still higher seed samples sizes. This fact suggests that experimental strategies should not primarily aim to investigate a maximum number of gene loci at the cost of sufficient sample size.

Acknowledgements

The technical assistance of G. Dinkel, M. Günther and R. Woelke is greatly appreciated. I also wish to thank M. Ziehe for valuable discussions, S. Franke and M. Waldschmidt for data processing, K. Radler for additional programming and E. Gillet for linguistic advice and M. Götemann for kindly typing the manuscript.

The seed probes were generously supplied by Bayer. Landesanstalt für forstliche Saat- und Pflanzenzucht. This study was financially supported by grants from the Deutsche Forschungsgemeinschaft, Bad Godesberg.

APPENDIX

Complete array of 10-locus genotypes of parental clones in seed orchard I (Schafknock) and/or seed orchard II (Herrgottschlag). Clone No. 1 is not present in I, clones No. 10, 28, 33, 35 not in II. Clones No. 29 and 30 were removed earlier from both seed orchards.

ENZYME GENE LOCUS	10-LOCUS GENOTYPES OF PARENTAL CLONES NO.																				
	1	2	3	4	5	6	7	8	9	10	11	12	13	14	15	16	17	18	19	20	21
GOT - A	22	22	22	22	22	22	22	22	22	22	22	22	22	22	22	22	22	22	22	22	22
GOT - B	35	33	23	55	35	35	55	33	23	55	33	55	35	35	35	55	55	55	55	33	35
GOT - C	44	24	24	44	24	24	24	44	24	22	24	24	44	44	24	22	24	44	44	24	44
LAP - B	22	22	22	22	02	22	22	22	22	12	22	22	22	22	22	22	22	22	22	22	22
MDH - A	13	33	33	33	33	33	13	33	33	33	33	33	33	33	33	33	33	33	33	33	33
MDH - B	11	11	11	11	11	11	11	11	11	11	11	11	11	11	11	11	14	11	11	11	11
MDH - C	33	34	33	34	34	34	34	34	33	34	34	34	33	33	34	34	34	33	33	33	34
MDH - D	24	44	44	02	24	24	24	24	[illegible]	11	12	24	24	22	44	24	24	12	22	24	14
SKDH - A	33	33	33	33	33	35	03	33	[illegible]	46	34	34	33	33	23	33	12	46	33	33	33
SKDH - B	33	33	33	33	33	33	33	33	33	33	33	33	33	33	33	33	33	33	33	33	33

	22	23	24	25	26	27	28	31	32	33	34	35	36	37	38	39	40	41	42	43	44	45
GOT - A	22	22	22	22	22	22	22	22	22	22	22	22	22	22	22	12	22	22	22	22	22	22
GOT - B	25	55	35	55	33	35	55	35	55	35	33	55	55	35	35	13	35	35	33	35	55	55
GOT - C	22	44	24	24	44	24	24	24	24	24	22	24	44	24	24	22	22	44	24	24	24	22
LAP - B	22	22	22	22	22	22	12	22	22	22	22	22	12	22	12	22	12	22	22	22	12	22
MDH - A	33	33	13	33	33	33	33	13	33	33	33	33	33	33	33	33	13	33	33	33	33	33
MDH - B	11	11	11	11	11	11	11	11	11	11	11	11	11	11	11	11	11	11	11	11	11	13
MDH - C	34	33	34	33	33	34	34	34	34	34	34	33	34	33	33	33	13	33	33	33	34	33
MDH - D	22	22	22	11	44	22	24	22	44	44	12	24	24	24	44	24	44	12	24	44	44	14
SKDH - A	33	33	33	34	33	33	33	33	34	23	33	03	33	33	33	34	33	33	33	35	33	03
SKDH - B	33	33	33	33	23	23	23	33	33	33	33	33	23	33	33	33	33	33	33	23	33	33

References

Adams, WT, Joly RJ (1980) Allozyme studies in loblolly pine seed orchards: Clonal variation and frequency of progeny due to self-fertilization. Silvae Genetica 29: 1-4

Chung, M-S (1981) Biochemical methods for determining population structure in Pinus sylvestris L. Acta Forest Fenn 173: 1-28

El-Kassaby (1981) Genetic interpretation of malate dehydrogenase isozymes in some conifer species. Journ Hered 72: 451-452

Friedman ST, Adams WT (1982) Genetic efficiency in loblolly pine seed orchards. In: Proc 16th South Forest Tree Improv Conf, Blacksburg, VA 1981. Polytech Inst State Univ, Publ 38: 213-224

Gregorius H-R, Ross MD (1981) Selection in plant populations of effectively infinite size: I. Realized genotypic fitnesses. Math Biosci 54: 291-307

Hadders G, Koski V (1975) Probability of inbreeding in seed orchards. In: Faulkner R (ed) seed orchards. For Comm Bull 54: 108-117

Jonsson A, Ekberg I, Eriksson G (1976) Flowering in a seed orchard of Pinus sylvestris L. Stud Forest Suecica 135: 1-38

Moran GF, Bell JC, Matheson AC (1980) The genetic structure and levels of inbreeding in a Pinus radiata D. Don seed orchard. Silvae Genetica 29: 190-193

Müller G (1976) A simple method of estimating rates of self-fertilization by analyzing isozymes in tree seeds. Silvae Genetica 25:15-17

Müller-Starck G (1979) Estimates of self- and cross-fertilization in a Scots pine seed orchard. In: Rudin D (ed) Proc Conf Biochem Genetics of Forest Trees, Umeå, Sweden 1978, Swed Univ Agric Sci, Report 1: 170-179

Müller-Starck G (1982a) Sexually asymmetric fertility selection and partial self-fertilization. 2. Clonal gametic contributions to the offspring of a Scots pine seed orchard. Silva Fennica 16(2): 99-106

Müller-Starck G (1982b) Reproductive systems in conifer seed orchards. I. Mating probabilities in a seed orchard of Pinus sylvestris L. Silvae Genetica 31: 188-197

Müller-Starck G, Ziehe M, Hattemer HH (1983) Reproductive systems in conifer seed orchards. 2. Reproductive selection monitored at an LAP gene locus in Pinus sylvestris L. Theor Appl Genet 65: 309-316

Müller-Starck G, Ziehe M (1984) Reproductive systems in conifer seed orchards. 3. Female and male fitness of individual clones realized in seeds of Pinus sylvestris L. Theor Appl Gen 68: in print

O'Malley DM, Allendorf FW, Blake GM (1979) Inheritance of isozyme variation and heterozygosity in Pinus ponebrosa. Biochem Gen 17: 233-250

Ross MD (1984) Frequency-dependent selection in hermaphrodites: the rule rather than the exception. Biol Journ Linn Soc 23: in print

Rudin D, Lindgren D (1977) Isozyme studies in seed orchards. Studia Forestalia Suecica 139: 1-23

Schmidtling RC (1983) Genetic variation in fruitfulness in a loblolly pine (Pinus taeda L.) seed orchard. Silvae Genetica 32: 76-80

Shen H-H, Rudin D, Lindgren D (1981) Study of the pollination pattern in a Scots pine seed orchard by means of isozyme analysis. Silvae Genetica 30: 7-15

Sweet GB (1975) Flowering and seed production. In: Faulkner R (ed) Seeds orchards. For Comm Bull 54: 72-82

Squillace AE, Goddard RE (1982) Selfing in clonal seed orchards of slash pine. Forest Sci 28: 71-78

Ziehe M, Gregorius H-R(1981) Deviations of genotypic structures from Hardy-Weinberg proportions under random mating and differential selection between the sexes. Genetics 98 (1), 215-230

Ziehe M (1982a) Sexually asymmetric fertility selection and partial self-fertilization. I. Population genetic impacts on the zygotic genotypic structure. Silva Fennica 16(2): 94-98

Ziehe M (1982b) Zygotic genotypic frequencies under selection on female or male gamete production in partially self-fertilizing plant populations. Heredity 49(3): 271-290

Multilocus analysis of external pollen contamination of a Scots pine (*Pinus sylvestris* L.) seed orchard

Kazutosi Nagasaka and Alfred E. Szmidt

Abstract

Genotypes at 18 allozyme loci were determined for all clones and their half-sib embryo families in a clonal seed orchard of Scots pine. The proportion of pollen gametes with genotypes that could not have been produced by parental clones was determined and used as a minimum estimate of external pollen contamination. We found that 37.8% of analysed embryos were fertilized by external pollen.

Introduction

The genetic consequences of mating patterns in forest tree seed orchards are of practical importance. The contamination of seed crops with non-orchard pollen is among the most important factors which can considerably deteriorate the genetic quality of seed orchard progenies. Information of this phenomenon is essential for sound orchard design and management. Allozyme markers have proved to be very useful for studying various factors affecting the genetic composition of seed orchard crops (e.g. Adams and Joly 1980, Müller-Starck 1982). The introduction of multilocus methods based on the simultaneous analysis of a large number of allozyme loci has furnished even more versatile means of such studies (e.g. Conkle and Adams 1977, Shaw and Allard 1981).

In this paper we present preliminary results of our studies of the degree of contamination with foreign pollen in a clonal seed orchard of Scots pine in Sweden.

Materials and methods

Cone collections were made in the Robertsfors clonal seed orchard of Scots pine in northern Sweden. The age of clones varied from 18 to 25 years. The total area of the orchard is 6.0ha. The actual number of clones in this orchard is 30. All cones were collected from one randomly chosen ramet from each clone. Seeds were extracted following kiln drying at app. 50°C. The seeds were supplied to us by Torbjörn Lestander of the Northern District Institute for Forest Improvement at Sävar. The multilocus genotypes of the parental clones and open-pollinated embryos were assessed at 18 allozyme loci, LAP-A, LAP-B, GOT-A, GOT-B, APH, FE, SDH-A, SDH-B, GDH, 6PGD-A, 6PGD-B, MDH-A, MDH-B, PGI-A, PGI-B, PGM-A, PGM-B, and ACO. Information concerning electrophoretic methods and formal genetics of these loci have been presented elsewhere (see Szmidt and Muona in this volume).

The parental clone genotypes were inferred from the allozyme patterns in haploid macrogametophyte tissue. Two clones produced only empty seeds and their genotypes were determined by analysing winter bud tissues. Twelve endosperms and embryos were analysed from each clone. All 30 parental clones could be individually identified based on their multilocus genotypes at 18 allozyme loci. The multilocus genotypes of pollen gametes originating from open-pollination were deduced from the comparison between allozyme patterns in macrogametophytes and the corresponding embryos. Due to poor resolution, genotypes of some embryos could not be scored at 18 loci. These embryos have been omitted from further analysis. Thus, the final results are based on multilocus data from 245 seeds in 28 clones. Those pollen gametes with genotypes that could not have been produced by any of the 30 parental clones were regarded as contamination from external pollen sources.

Results and discussion

The multilocus estimates of contamination with pollen from external sources are given in Table 1. There was much variation among individual clones with regard to the rate of contamination. However, the number of embryos analysed in individual clones was too small to permit a more detailed analysis of these differences.

Table 1. Proportion of embryos fertilized by external pollen

clone number	number of seeds sampled	rate of contami-nation	clone number	number of seeds sampled	rate of contami-nation
1.AC1013	9	0.333	16.AC3010	11	0.273
2.AC1075	7	0.571	17.AC3023	10	0.500
3.AC1077	8	0.750	18.AC3025	11	0.545
4.AC1978	10	0.100	19.AC3031	2	0.000
5.AC2035	10	0.300	20.AC3046	7	0.571
6.AC2077	5	0.800	21.AC3055	8	0.375
7.AC2084	7	0.286	22.AC3058	--	-----
8.AC2087	6	0.500	23.AC3063	11	0.182
9.AC2091	8	0.500	24.AC3065	11	0.455
10.AC2095	5	0.200	25.AC3066	5	0.200
11.AC2101	10	0.200	26.AC4208	10	0.600
12.AC2103	--	-----	27.AC4214	12	0.500
13.AC2106	11	0.364	28. Y2037	12	0.250
14.AC3008	9	0.667	29. Y3014	9	0.556
15.AC3009	9	0.333	30. Y3088	12	0.417
total				245	0.378

The average level of pollen contamination over all embryos analysed was estimated to be 37.8%. Considering that the investigated seed orchard is already 18 to 25 years old and in full seed production, this high contamination with external pollen is somewhat surprising. In addition our present estimate must be regarded as minimum. As mentioned earlier, only those pollen gametes with genotypes that could not have been produced by orchard clones were regarded as "foreign". However, it is likely that Scots pine stands surrounding Robertsfors also produce gametes with genotypes matching those generated by orchard clones. These gametes cannot be distinguished from orchard pollen and would go undetected.

To estimate the expected proportion of such indistinguishable gametes the surrounding stands must be determined(Smith and Adams 1983). This has not been done in this study. However, we have already collected such materials and intend to analyse them the nearest future. Very few estimates of contamination with external pollen in forest tree seed orchards are available so far. Friedman and Adams (1982) studied contamination in a 16 year old seed orchard of Pinus taeda surrounded by a 122m pollen dilution zone. They found that 28% of seeds received pollen from the surrounding stands. Up to 43% of contamination with external pollen was found in a seed orchard of Pseudotsuga menziesii (Smith and Adams 1983). Heavy seed crops often produced by young Scots pine orchards where only little orchard pollen is available, is another indication of considerable impact of foreign pollen sources. Scots pine pollen is equipped with air sacks which retard falling velocity and facilitate dispersion (Stanley and Linskens 1974). Using computer simulations Gleaves (1973) has demonstrated that there is more pollen flow into regularly spaced populations than into those where trees are randomly distributed. In contrast to natural stands where plants are usually scattered in an irregular manner, seed orchard clones are arranged in regular widely spaced rows. Thus they may represent an easy object for penetration of pollen clouds approaching from neighbouring stands.

Our results indicate that even in old seed orchards, contamination with foreign pollen may be high. There is a narrow pollen dilution zone (consisting of birch, 10m wide, 3-5m high) around Robertsfors. It appears that it is not very effective in preventing foreign pollen dispersal into this seed orchard. The nearest Scots pine stands to the south and east of Robertsfors are separated by 500-800m of open flat field. In addition, one small stand of 3ha including big old Scots pine trees is located very close to the northern border of this orchard and is separated by only pollen dilution zone. Apparently, pollen clouds produced by these stands may enter this orchard relatively easily.

One possible factor which could raise our estimates of contamination in Robertsfors is the existence of foreign clones in the orchard. As practice shows, this is not uncommon in forest seed orchards. Only one ramet per clone was tested in our study. Assuming that "illegitimate" clones are present in Robertsfors, it is likely that they produce gametes other than those generated by the parental clones tested and thus would be regarded as "foreign".

Our present results concern only a single annual seed crop. Observations in Scots pine seed orchards in Sweden indicate large temporal variation in flowering and seed production. Varying amounts of pollen produced within a seed orchard are likely to affect this contamination rate. Little is known, however, of the temporal aspect of this process. It would also be of interest to know whether certain parts of a seed orchard are more endangered by the foreign pollen inflow than others e.g. central versus marginal parts. Such informations could be helpful in proper seed collection and management of the existing seed orchards. More detailed studies are required in order to elucidate these problems. Such studies are already conducted in our laboratories and their results will be presented elsewhere.

Acknowledgements

We wish to thank Prof. Dag Lindgren and Dr. Outi Muona for helpful comments. Maj-Len Åman helped us with kind technical assistance. Seed samples were supplied by Torbjörn Lestander of the Institute for Forest Improvement at Sävar. This study was financially supported by grant from the Cellulose Industries Council for Technology and Forest Research.

Literature

Adams,W.T. and Joly,R.J. 1980. Allozyme studies in Loblolly pine seed orchards: clonal variation and frequency of progeny due to self-fertilization. Silvae Genet. 29: 1-4.

Conkle,M.T. and Adams,W.T. 1977. Use of isoenzyme techniques in forest genetic research. Proc. 14th South. Forest Tree Improvement Conf. Gainesville, Florida.

Friedman,S.T. and Adams,W.T. 1982. Genetic efficiency in Loblolly pine seed orchards. Proc. 16th South. Forest Tree Improvement Conf. Blacksburg, Virginia. 213-224.

Gleaves,J.T. 1973. Gene flow mediated by wind-borne pollen. Heredity 31: 355-366.

Müller-Starck,G. 1982. Reproductive systems in conifer seed orchards. I. Mating probabilities in a seed orchard of Pinus sylvestris L. Silvae Genet. 31: 5-6.

Shaw,D.V. and Allard,R.W. 1981. Analysis of mating system parameters and population structure in Douglas-fir using single-locus and multilocus methods. Proc. of symposium on Isozymes of North American Forest Trees and Insects. Berkeley, California. 18-22.

Smith,D.B. and Adams,W.T. 1983. Measuring pollen contamination in clonal seed orchards with the aid of genetic markers. Proc. 17th South. Forest Tree Improvement Conf. Athens, Georgia. 69-77.

Stanley,R.G. and Linskens,H.F. 1974. Pollen, Biology, Biochemistry, Management. Springer Verlag, Berlin, Heidelberg, New York.

GENE DISPERSION AND SELFING FREQUENCY IN A SEED-TREE STAND OF *PINUS SYLVESTRIS* (L.)

R. Yazdani, D. Lindgren and D. Rudin

Abstract

The spatial distribution of unique allozymes among 785 naturally regenerated plants (aged 10 - 20 years) established under seed trees was determined using electrophoretic techniques. Genes from four seed trees carrying marker alleles were traced in seedlings in the vicinity of those trees. The contribution of a seed tree to plant regeneration close to that seed tree was about 5 %; thus plants are usually not the progeny of the closest seed tree. Groups consisting mostly of half-sibs were not evident. As the half sibs seemed to be distributed rather widely, half-sib mating is probably not a common source of inbreeding in a naturally regenerated stand.

A model of gene dispersal was developed. According to this model the proportion of all genes dispersed at certain distances from the source are calculated. The approximate percentages of genes travelling different distances are 0 - 5 m, 4 percent; 5 - 10 m, 8 percent; 10 -15 m, 11 percent; 15 -50 m, 40 - 75 percent.

Examination of embryos from seeds produced by seed trees has revealed a large excess of homozygotes. This excess of homozygotes probably resulted from inbreeding caused by selfing. Estimation of selfing by rare and unique isozyme markers from thirteen trees gave an average value of 11.8 percent at the embryo stage. Selfing rates appear to be dependent on seed-tree density, climate conditions at time of pollen flight and size of pollen and seed crop.

Introduction

A considerable part of Scots pine (*Pinus sylvestris* L.) reforestation in Sweden is carried out by means of natural regeneration with seed trees. Natural regeneration under seed trees is regarded as satisfactory when a sufficient number of developing plants are established per hectare, and these plants are evenly distributed throughout the stand. In southern Sweden, the longest acceptable time for satisfactory natural regeneration after final felling is five years, and in other parts of Sweden ten years, except for some interior parts of Norrland, where it is around fifteen years. The success of natural regeneration is strongly dependent on climatic conditions, latitude and elevation of the origin (Elfving, 1978).

In southern Sweden, Scots pine produces an average of 5,000 viable seeds per tree per year, while in the north and at higher elevation the production amounts to only a few hundred viable seeds per tree per year (Hagner, 1965). In northern Sweden the seeds of Scots pine are released during April - July, with maximum release at the end of May. Some factors of importance in connection with seed dispersal are wind velocity, air turbulence, fall height and seed-falling rates (Wilhelm-Schmidt, 1918, 1925 and Sarvas, 1962). Long-winged and short-winged seeds are also transferred to different extents (Hesselman, 1934). The large proportion of seeds are spread not more than 10 - 15 m from the seed trees. With strong wind, seeds may be transferred 20 - 30 m in the wind direction. The lighter seeds, which are carried longer distances by the wind, usually do not have a good germination capacity. Fewer than 2 % of germinable seed typically reach seedling size in virgin ground cover (Hagner, 1965). While environmental factors greatly influence the seed germination capacity, genetic components seem to be important in whether germinated seeds will continue growth to plant stage (Rudin, et al., 1977). Conifer display a high degree of inbreeding depression, the avoidance of inbreeding is based on elimination of inbred individuals after fertilization (Tigerstedt et al. 1982, Yazdani et al. 1984). For Douglas fir, Campbell (1979) estimated that well over 20,000 seeds are produced to replace one tree in an older growing stand. According to Campbell (1979), genotypes surviving to the seedling stage are strongly screened by the physical environment. The seedling population surviving to adult stage at a particular site can also undergo genetic shifts as a result of local adaptation (Bradshaw, 1972). It is likely that much of the variation within populations is adaptive (Kleinschmidt, 1979; Campbell, 1979).

The objectives of the present work were to:

a. Determine the pattern of gene dispersal from seed trees into the established reproduction under the seed trees.
b. Estimate selfing frequency in the seed-tree stand.

Material and methods

Stand data

The study was carried out in a 100-year-old Scots pine (Pinus sylvestris L.) seed-tree stand at Svartberget, close to Vindeln, in northern Sweden (lat. 64° 14' N, long. 19° 47' E and elev. 200 m). The number of seed trees per hectare in the test plot was 121. The seed-tree stand was established in 1955. This seed-tree stand was closely surrounded by other mature stands of Scots pine. The seed trees were removed in the autumn of 1979 and the stumps of individual seed trees were marked with metal identification numbers. The position of the seed trees is indicated in

Figure 1. The number of plants regenerated under the seed trees was estimated to be 5,200 plants per hectare in 1983.

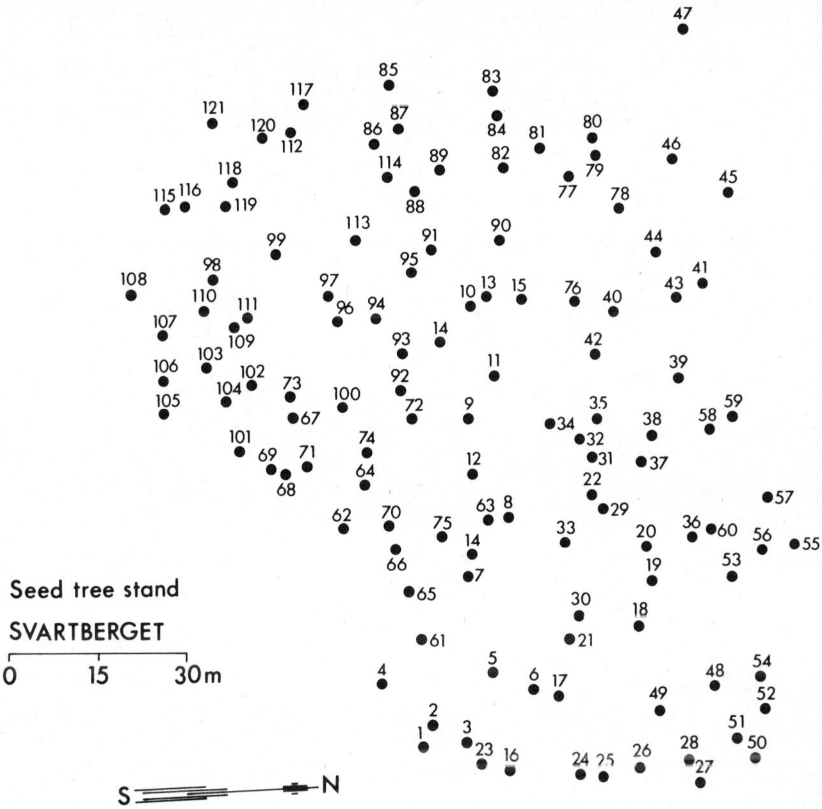

Figure 1.The position of seed trees within the seed-tree stand at Svartberget. The numbers correspond to seed trees.

Collected material

Wind-pollinated cones were collected from all seed trees in 1979. Buds were collected from 785 naturally-regenerated seedlings between the estimated ages of ten and twenty years at distances of 0 - 5, 5 - 10 and 10 - 15 meters from eight marker trees with rare or unique isozyme markers and in directions north, south, east and west (cf. Figure. 2).

For isozyme analysis, bud tissue on seedlings and seed tissue on seed trees were used. Isozyme markers in four trees were found to be unique for the whole seed-tree stand. These trees are indicated by the numbers 44, 67, 83 and 98, and by the following marker alleles: LAP-B1, LAP-A1, GOT-A1 and LAP-B01 (Figure. 2).

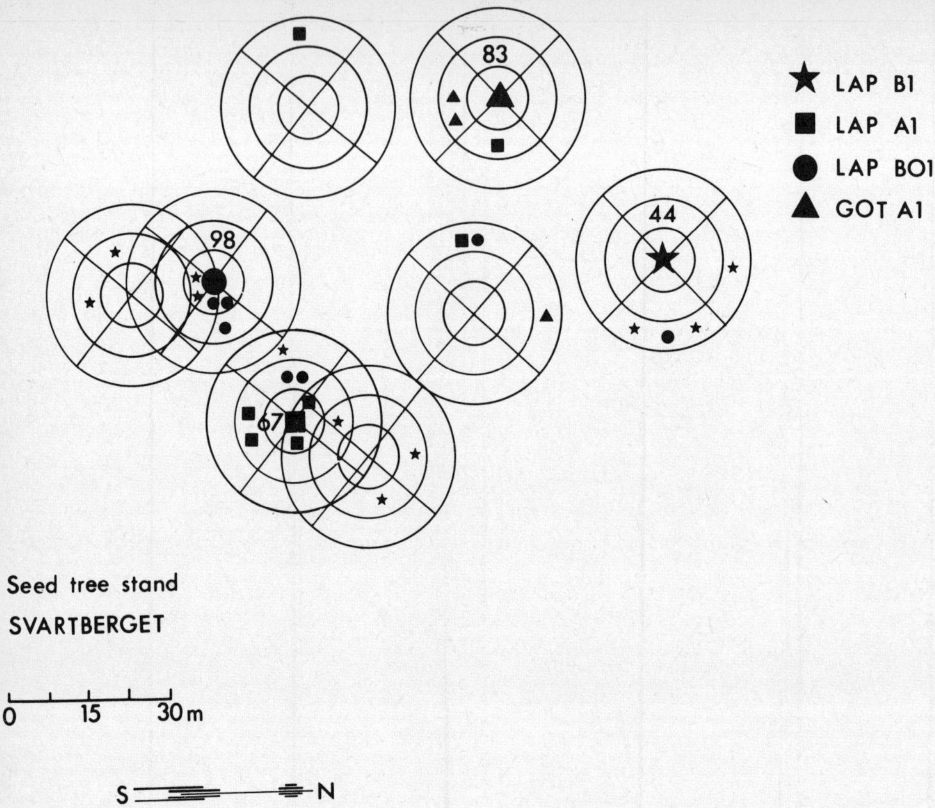

Figure 2. The position of the four marker trees (44, 67, 83 and 98) within the seed tree stand at Svartberget. Circles and lines indicate the distribution of test material collected from naturally regenerated seedlings. About 10 seedlings were sampled within each sector. Small symbols show the distribution of seedlings carrying the same isozyme markers as the marker seed trees, and thus expected to be progeny of those trees.

For estimation of selfing frequency and the inheritance pattern of enzymes the seeds from twelve parent seed trees with either rare or unique isozyme markers were analysed. For further information on material and methods used for the analyses, see Yazdani, et al. (1984).

Model for gene dispersal analysis

In this investigation the proportion of marker seedlings growing at an area (a spot) is measured. If an analysed spot is situated at a larger distance from the marker tree, the same proportion means a larger total contribution. The dispersal at large distances is not possible to measure, but the total dispersal must still be estimated to analyse the fraction at different distances. That some of the parents to the analysed seedlings are other trees than the analysed seed trees is a problem still to be dealt with.

The model is given here in a specific form to fit the problem without too complicated terminology. However, it may easily be generalized.

Five different alternative intervals for gene dispersal are considered.

The following denotations are introduced to denote the alternatives:

N total number of analysed seedlings

M number of marker seedlings

Q the measured proportion of seedlings carrying a marker on a certain spot at a certain distance (M/N at the spot).

F(a,b) is the proportion of genes from the marker tree in seedlings at distances between a and b from the marker tree. F is calculated from M and N by subtracting the influence of the background.

G(a,b) is the proportion of all genes dispersed ending up at distances between a and b from the source. (This is desirable to know but difficult to measure, especially for large distances).

The following denotations will be used for the five different alternatives

		Proportion marker seedlings	
		Corrected for background	Dispersed proportion
0 - 5	m from the source	F (0,5)	G (0,5)
5 - 10	m from the source	F (5,10)	G (5,10)
10 - 15	m from the source	F (10,15)	G (10,15)
> 15	m from the source, but within the seed tree stand area	F (>15)	G (>15)

C is the proportion of genes dispersed outside or originating from outside the seed-tree stand ("Background").

Based on the definition:

$G\ (0,5) + G\ (5,10) + G\ (10,15) + G\ (>15) + C = 1$, thus $C = 1 - \Sigma G$.

The "gene density" in this scale will be d = density of seed trees = 121 trees per hectare = 0.0121 trees/m^2 = 0.0121 "gene dispersal units" per m^2.

G may be calculated by multiplying F with the area and gene density.

$$G(a,b) = \Pi(b^2 - a^2)d\,F(a,b) \qquad (1)$$
$$\Pi(b^2 - a^2)d = 0.950;\ 2.851;\ 4.751$$
for resp $(a, b) = (0,5); (5, 10); (10, 15)$ m.

A reciprocity may be postulated. The distribution of genes travelling different distances from a point source is (according to the model) expected to be the same as the distribution of distances from the source for genes reaching a certain spot. Thus as many genes leave a specified seed tree area as those arriving to it. The fraction C may therefore be regarded as the fraction of genes not originating in the seed-tree stand.

The "background" C contribution is assumed to be the same everywhere in the seed-tree stand. To correct for the background influence, the contribution 2CpN (where p = gene frequency of the marker in the background) is deducted from the number of marker seedlings and the number of analysed seedlings is reduced to N(1-C). Thus:

$$F = (M - 2CpN)/N(1 - C). \qquad (2)$$

F-values pooled over several systems were calculated by summing numerators and denominators separately over the systems and dividing the summed values. The opposite procedure would give trouble with negative numerators and a large statistical error caused by small denominators. Thus:

$$F = \Sigma\,(M - 2C_pN)/\,\Sigma\,N(1 - C) \qquad (3)$$

G (>15) is calculated as a difference, as it seems risky to assume that F (>15) is constant over the whole area.

Results and discussion

Inheritance check

The inheritance of the allozymes for certain allelic combinations was determined for different loci by using haploid megagametophytes from 100 seeds obtained from single trees (Table. 4). Segregation ratios were close to 1:1 as expected for co-dominant alleles at all heterozygous loci, except for the GOT-B locus with allelic combination 1 - 3. As similar distortions were found in two trees (10 and 108) it is probable that the distortion is an inherited phenomenon.

Gene dispersion through seeds and pollen

To study the actual pattern of gene dispersion in a young population originating from individual parent trees, all isozyme markers from parent and offspring were matched. Where other isozymes excluded parenthood by the marker trees, data from these individuals were excluded from the analyses. The position of possible parents and the distribution of seedlings carrying parental marker alleles is demonstrated in Figure 2. There may also be markers originating from trees outside the stand. This is especially probable for LAP-B1, as it was found in the pollen cloud at rather large distances from marker tree no. 44 (Table 4). Therefore, it seems probable that the LAP-B1-carrying seedlings at a large distance from the marker tree are not progeny from that tree. If this type of work is continued, it is recommended that a supporting check of parentage with other loci considered. There are evident groups of marker seedlings close to trees 83, 67 and 98, probably constituting half- or full-sibs, with the neighbouring marker tree as a parent (maternal or paternal). There is an impression that some marker seedlings occur in clusters in the same direction from the supposed parent rather than being spread at random in all directions. This may reflect an influence of wind direction during pollen and seed dispersal.

In Figure 3 the percentage of seedlings carrying a marker allele is demonstrated as a function of the distance from the supposed marker parent. Since the marker trees are heterozygous, only half of their progeny will carry the markers, and the percentage has to be doubled to arrive at an estimate of the frequency of progeny originating from a marker tree. It seems that somewhat less than 9 percent of the seedlings have the closest seed tree as a parent (i.e. somewhat more than 90 percent do not have the closest tree as one of their parents). This seems to be an important finding for understanding the genetic structure of a naturally regenerated forest.

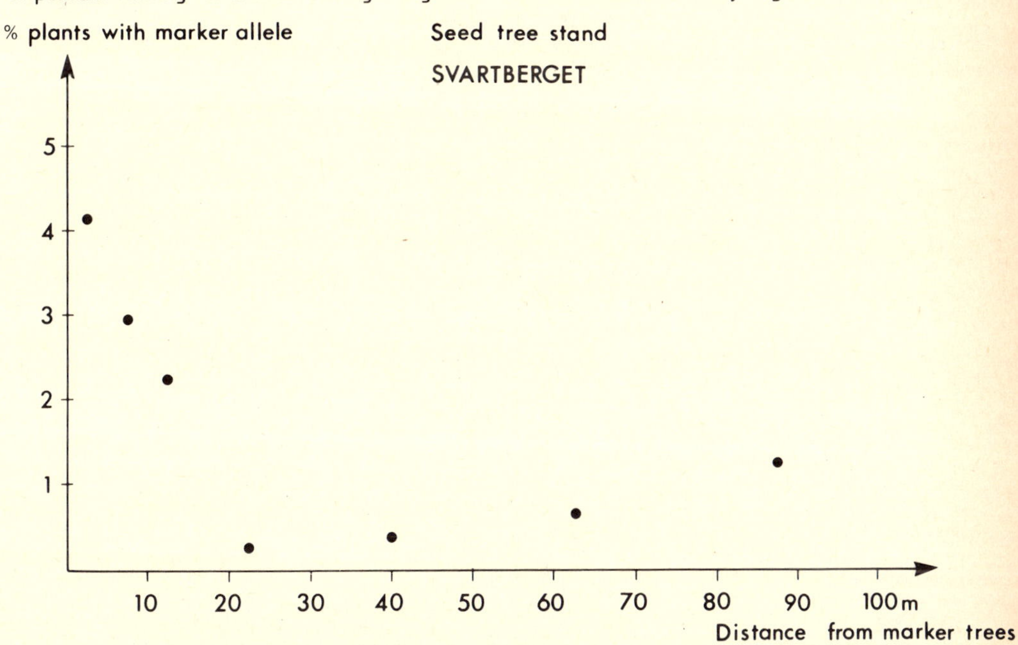

Figure 3. The percentage of seedlings carrying marker alleles at different distances from the supposed marker parents.

It has sometimes been suggested that a natural forest may be built up of small groups of full or half-sib families, and that inbreeding due to fertilization between sibs will be common. The pattern obtained in this study (Figure 2) does not seem to support such a theory. Tigerstedt et al. (1982) concluded that regeneration groups around old growth trees also do not appear to be particularly related to that tree.

The distribution of seedlings carrying markers is presented in Table 1 as a function of the distance to the (closest) marker tree. The entity Q, which is the measured proportion of seedlings carrying a marker on a certain spot at a certain distance, is easily calculated as the quotient between the values in Table 1.

The pattern is a result of pollen flight as well as seed dispersal from the parent trees. There is also a "background" of markers originating from trees outside the stand, which might also be genes from trees cut away when the seed-tree stand was established.

The marker seedlings within the 15 m circle from the marker tree usually seem to be associated with that tree. At longer distances we found the pattern difficult to interpret, and certainly distorted by the background. Even just a few progeny seedlings identified at a long distance from the marker tree may reflect that a large proportion of genes are transmitted long distances. Furthermore, the sampling was done in a symmetric way only within 15 m from the marker trees. Thus in Table 2 and 3 all marker seedlings more distant than 15 m from marker trees are pooled.

Table 1. Frequency of seedlings with marker genes on different distances from the (closest) marker tree/all analysed seedlings. The quotient = Q.

Marker tree	Marker	Distance to seedlings (m) 0 - 5	5 - 10	10 - 15	15 <
44	LAP B1	0/10	0/20	3/66	8/686
67	LAP A1	2/26	2/46	0/28	3/682
83	GOT A1	0/15	2/46	0/39	1/682
98	LAP BO1	2/45	1/45	0/0	4/682
	Pooled	4/96	5/167	3/133	16/2732
	Q	.042	.030	.024	.006
10	GOT B1	2[1)]/17	3/26	3[1)]/43	
108	GOT B1	2/54	2/49	0/0	19/393
112[2)]	GOT B1/B1	2/21	2/21	3/60	
74	F-EST AO1	0/31	0/44	2/20	10/585
117	F-EST AO1	0/21	0/21	1/60	
	Pooled	6/144	7/161	9/183	29/978
	Q	.042	.043	.049	0.30

1) One B1/B1 counted twice
2) The marker tree is homozygous for the markers, thus all progeny will carry the marker

Table 2 Values for estimation of background

Markers	Markers among seed trees	Q(> 15) (Table 1)	Allele frequency of marker in background (Gullberg et al. 1984) p	Q (> 15) /2 p
LAP B1	1	.0117	.0177	.33
LAP A1	1	.0044	.0109	.20
GOT A1	1	.0015	.0050	.15
LAP BO1	1	.0059	.0087	.34
GOT B1	4	.0483	.0254	.64
F-EST A01	2	.0171	.0014	(>1)

Correction for background

Gullberg et al (1984) investigated the gene frequencies in 7 Scots pine populations in middle Sweden. They concluded that the allele frequencies in isozyme systems show little geographic variation. Therefore, it seems reasonable to assume that the allele frequencies in the "background" genes are similar to the mean values in the above-mentioned populations. Those are listed in Table 2. It may be noted that there seems to be a relation between marker-seedling frequency outside the 15 m radius Q (>15) and the expected background P, indicating a partly outside origin of these seedlings. However, the F-EST A 01 carriers do not fit into this pattern. It seems also unreasonable to get two independent F-EST A 01 if the gene frequency is as low as indicated. However, the trees may be related, and further relatives may contribute to the local gene frequency.

The contribution to the proportion of marker-seedlings from the background will be 2 Cp, where C is proportion of background and p allele frequency in background. If C exceeds Q (>15)/2p (Table 2, last column), the proportion of marker seedlings originating from background alone would exceed what is found outside the 15 m radius. The Q (> 15) values for all four single markers indicate a C < 0.35. However, this upper bond assumes no marker seedlings from the marker tree detected outside the 15 m radius, and this seems unreasonable. Thus the upper bond for the probable interval is set to C = 0,30. If C = O and the genes are equally distributed in the area, a frequency of 1/121 = 0.0083 (or lower) would be expected (4 x 0.0083 for GOT B1 and 2 x 0.0083 for F-EST A 01). This is the magnitude found, and thus C = O is not unreasonable. Thus it is reasonable that 0 < C < 0.30.

As the exact value of the background is highly uncertain, assumptions between C = O and 0.4 were tested (Table 3). The expected proportion of genes dispersed different at distances from the seed tree (G) was computed for unique markers, and the results presented in Table 3. F was calculated by Formula (3), inserting M and N for Table 1, C as assumed and P from Table 2. The obtained F is then utilized for calculating G by Formula (1).

The resolution of the pattern for the non-unique markers(GOT B 1 and F-EST AO1, Table 1) seems unsatisfactory for a more detailed analyses, as Q (>15) is not much lower than Q at closer distances to the marker trees. This may be caused by rather high frequency of the marker alleles in the seed-tree stand as well as in the local "background". The pattern found for non-unique markers is not in strong contradiction to that of unique markers. There is, however, a clear tendency that less marker seedlings are found within the 15 m radius compared to >15 m for non-unique compared to the unique markers (Table 1). It seems therefore reasonable to adjust the estimates G (0,5) and G (5,10) obtained from unique markers (Table 3) a little downwards. It is somewhat surprising that no F-EST A01 seedlings were found within 10 m from the marker trees, but this may be the result of untypical situations of flowering or wind and other conditions at critical moments.

G (0-5), G (5-10) and G (10-15) in Table 3 are remarkably little dependent on C. Irrespective of the C-valuet, it seems possible to conclude that the genes typically travel the following distances:

Table 3 Proportion of genes dispersed at different distances. Pooled data on the four unique marker trees with different assumptions of background.

		Distance from markers (m)			
Calculated entity	Assumed C	0 - 5	5 - 10	10 - 15	15 <
F	0	.0417	.0299	.0239	.0059
F	0.1	.0441	.0311	.0223	.0042
F	0.2	.0473	.0328	.0219	.0020
F	0.3	.0512	.0347	.0215	(< 0)
G	0	.0402	.0866	.1283	.7449
G	0.1	.0426	.0900	.1076	.6598
G	0.2	.0456	.0950	.1057	.5537
G	0.3	.0494	.1005	.1037	.4464

Distance travelled	Share of genes
0 - 5 m	4 percent
5 - 10 m	8 percent
10 - 15 m	11 percent
15 - 50 m	40 - 75 percent
> 50 m	0 - 35 percent

If judged reliable and typical, these results may be very useful for modelling the genetic structure of populations of forest trees. However, first it should be checked if a similar distribution occurs in other seed-tree stands.

Estimated selfing frequency in the seed-tree stand

By studying embryos and megagametophytes of the same seed one can gain information about the genotypic composition of the individual mother tree as well as the gene composition of the fertilizating pollen cloud. This allows us to calculate the gene frequency in tho paternal population and determine the rate of selfing in the embryos.

Rare and unique isozyme markers from thirteen trees were used for studying the selfing frequency in thirteen seed parents from the seed-tree stand (Table 4). All thirteen marker trees were heterozygous for either unique or rare alleles, with gene frequencies in the seed-tree stand between 0.005 and 0.04. Selfing was estimated by looking at the frequency of the rare or unique allele in the pollen participating in fertilization of zygotes from the same tree, and multiplying it by 2. As there may be marker pollen of other origin present, this procedure slightly overestimates the amount of selfing. There seem to be striking differences in the frequency of selfing among seed trees. The percentage of selfing in the embryo stage is estimated at between 4 and 24 %, with an average selfing of 11.8 percent for the thirteen trees (lower and upper 95 % confidence limit 7.10 and 16.44, respectively). Since the isozyme patterns were not easy to determine in many seeds of poor quality, the true value may be higher, but on the other hand these poor seeds may not have resulted in plants. The estimated natural selfing frequency in different coniferous species seems to vary from 2 to 40 percent (Squillace, 1974).

An average selfing frequency of 24 percent was discovered by studying progeny of individual trees via embryos from seven trees from the seed-tree stand at Gårdtjärn in northern Sweden, with a density of approximately 12 - 18 trees per hectare (Rudin, et al., 1977). The average selfing in two-year-old half-sib seedlings originating from four seed trees in the same stand was approximately 17 percent. It is likely that the lower value of selfing found in our study, as compared to the value in the study mentioned above, is due to the higher density in our stand. If this is the case, it should be accepted that by spacing the seed trees widely we may increase the

Table 4. Estimation of selfing frequency with the help of rare and unique isozyme markers. Continuation on next page.

Tree no	Locus	Genotype	Allele segregation	Gene frequency in pollen cloud		Selfing frequency %
10	GOT-	B1/B3	39/23	B1	0.05	
				B2, B18, B22	0.43	10
				B3	0.46	
				X	0.06	
14	MDH	B1-0/B2-4	55/41	B1-0	0.04	
				B1-3	0.51	-
				B2-4	0.25	
				Blank	0.20	
				X	0	
33	LAP	B2/B3	55/45	B1	0.02	
				B2	0.87	18
				B3	0.09	
				X	0.02	
44	LAP	B1/B3		B1	0.08	
				B2	0.55	16
				B3	0.19	
				X	0.18	
57	MDH	A1/A2	51/49	A1	0.07	
				A2	0.92	14
				X	0.01	
57	MDH	B1-3/B1-3		B1-3	0.76	
				B2-4	0.22	-
				X	0.02	
67	LAP	A1/A2	47/53	A1	0.12	
				A2	0.83	24
				A3	0.01	
				X	0.04	
71	MDH	A1/A2	50/50	A1	0.06	
				A2	0.89	12
				X	0.05	
71	MDH	B2-4/B2-4		B1-3	0.56	
				B2-4	0.43	-
				X	0.01	
74	F-EST	A01-A1	51/49	A01	0.02	
				A1	0.74	
				A2	0.04	4
				A3	0.12	
				X	0.08	

74	LAP	A2/A3	52/48	A2	0.93	
				A3	0.05	10
				X	0.02	
74	LAP	B2/B2	-	B01	0.01	
				B1	0.02	-
				B2	0.94	
				B3	0.03	
				X	0	
83	GOT	A1/A2	52/48	A1	0.05	
				A2	0.91	10
				X	0.04	
83	GOT	B18/B18	-	B1	0.01	
				B18	0.22	
				B2, B22	0.32	-
				B3	0.38	
				X	0.07	
98	LAP	B01/B2	56/44	B01	0.02	
				B2	0.91	4
				B3	0.04	
				X	0.03	
108	GOT	B1/B3	68/32	B1	0.06	
				B2, B18, B22	0.39	12
				B3	0.50	
				X	0.05	
112	GOT	B1/B1	-	B1	0.11	
				B2, B18, B22	0.39	11
				B3	0.43	
				X	0.07	
117	F-EST	A01/A3	49/50	A01	0.04	
				A1	0.58	
				A2	0.07	8
				A3	0.19	
				X	0.12	
117	LAP	A2/A2	-	A2	1.00	-
				X	0	
117	LAP	B2/B2	-	B1	0.03	
				B2	0.95	-
				B3	0.02	
				X	0	
					Mean:	11.8 %

X = Unidentified alleles

level of spontaneous selfing. Since selfing leads to increased mortality and a decrease in wood production in the stand (Franklin, 1969; Koski, 1971, 1973; Forshell, 1974 and others), some restrictions should be enforced against too wide a spacing. The average selfing reported for a clonal seed orchard of P. sylvestris at Östteg (Umeå), located 60 kilometers from our seed-tree stand, was approximately 6 percent (Shen, et al., 1981), and similar selfing frequencies have been found in other studies in seed orchards (Rudin and Lindgren, 1977).

Genetic structure between three different life stages in a naturally regenerating stand of Pinus sylvestris has been studied by Yazdani et al. (1984). In this study, the embryo population contained an excess of homozygotes while young and adult populations were found to fit panmictic expectations concerning their genotypic structure. It is suggested by Tigerstedt et al. (1982), that selection against high proportions of inbreds in seedling populations is important to avoid inbreeding depression in natural populations.

On the basis of the data, one can assume that the seeds produced or collected in seed-tree stands in comparison to those from a seed orchard consist of more selfed progeny and are therefore of somewhat inferior quality. Many selfed plants in a natural regeneration will die, leaving more space to others, and causing little loss in production. However, on large parts of the naturally regenerated area there is not a sufficient number of plants, and inbred plants may cause a considerable loss in production. The results discussed here are valid for this particular stand with its particular climate and density of seed trees. It may be of interest to repeat the study over a range of seed-tree stands with various climatic conditions and stand densities.

Acknowledgements

We are most grateful to the Swedish Council For Forestry and Agricultural Research, which has given financial support to this study. We are grateful to Evert Jeansson, who provided valuable comments during the preparation of the manuscript. We thank Lars-Göran Lejdebro and Marja-Leena Lejdebro for seed collection and seed extraction of experimental material. To Karin Ljung, Barbro Harbom and Gun Lindkvist many thanks for skillful technical assistance. Ulla Nyström, Helene Risberg and Gertrud Lestander are warmly thanked for illustrating and typing the manuscript.

Literaturereferences

Bradshaw, A.D. 1972. Some of the evoutionary consequences of being a plant. Evol. Biol. 5: 25 - 47.

Campbell, R.K. 1979. Genecology of douglas-fir in a watershed in the Oregon cascades. Ecology 60: 1036 - 1050.

Elfving, B. 1978. Återväxt efter bränning under fröträd. Hugin-rapport nr 10. Sveriges lantbruksuniversitet, Inst för skogsskötsel, 901 83 Umeå, Sweden.

Forshell, C.P. 1974. Seed development after self-pollination and cross-pllination of Scots pine, Pinus sylvestris L. Studia Forestalia Suecica nr 118, 37 pp.

Franklin, E.C. 1969. Inbreeding depression in metrical traits of Loblolly pine (Pinus taeda L.). As a result of self-pollination. Technical report No. 40, School of Forest Resources, North Carolina State University.

Gullberg, U., Yazdani, R., Rudin, D. and Ryman, N. Allozyme variation in Scots pine (Pinus sylvestris L.) native to Sweden, (in press).

Hagner, S. 1965. Yield of seed, choice of seed trees, and seed establishment in experiment with natural regeneration. Studia Forestalia Suecica nr 27: 1 - 43.

Hesselman, H. 1934. Några studier över förspridningen hos gran och tall och kalhyggets besåning. Meddelanden från Statens Skogsförsöksanstalt, Häfte 27, Nr. 5.

Kleinschmit, J. 1979. Limitations for restriction of the genetic varation. Silvae Genetica 28: 61 - 67.

Koski, V. 1971. Embryonic lethals of Picea abies and P. sylvestris L. Studia Forestalia Fenniae 75, 3, Helsinki.

Koski, V. 1973. On self-pollination, genetic load, and subsequent inbreeding in some conifers. Communications Instituti Forestalis Fenniae 78, 10, Helsinki.

Rudin, D. 1975. Inheritance of glutamate-oxalate-transaminase (GOT) from needles and endosperms of Pinus sylvestris. Hereditas 80: 296 - 300.

Rudin, D., Eriksson, G. and Rasmuson, M. 1977. Inbreeding in a seed tree stand of Pinus sylvestris L., in northern Sweden. A study by the aid of the isozyme technique. Research Notes by Department of Forest Genetics nr 25.

Rudin, D. and Lindgren, D. 1977. Isozyme studies in seed orchards. Studia Forestalia Suecica Nr 139, 23 pp.

Sarvas, R. 1962. Investigations on the flowering and seed crop of Pinus sylvestris. Comm. Inst. Forest. Fenn. 53, 4, 198 pp.

Schmidt, W. 1918. Die Verbreitung von Samen und Blutenstaub durch die Luftbewegung. Österreichische Botanische Zeihschrift Jahr 1918.

Schmidt, W. 1925. Der Massenaustausch in freier Luft and Verwandte Erscheinungen. Probleme der Geophysik VII. Hamburg.

Shen, H-H., Rudin, D. and Lindgren, D. 1981. Study of the pollination pattern in a Scots pine seed orchard by means of isozyme analyses. Silvae Genetica 30: 7 -15.

Squillace, A.E. 1974. Average genetic correlations among offspring from open-pollinated forest trees. Silvae Genetica 23, 149 - 156.

Tigerstedt, P.M.A., Rudin, D., Niemelä, T. and Tammisola, J. 1982. Competition and neighbouring effect in a naturally regenerating population of Scots pine. Silva Fennica. 16: No. 2, 122 - 129.

Yazdani, R., Muona, O., Rudin, D. and Szmidt, A.E. Genetic structure of a seed-tree stand and naturally regenerated plants of Pinus sylvestris L. (In press).

GENETIC CONSTRAINTS ON THE EVOLUTION OF PLANT REPRODUCTIVE SYSTEMS

Brian Charlesworth

ABSTRACT

Processes involved in the evolution of plant reproductive systems are reviewed. It is shown that constraints imposed by the rules of genetic transmission, and by the properties of available genetic variation in the breeding system, may significantly affect the outcome of natural selection on reproductive systems. Theoretical predictions of selection models are compared with the evidence from genetic and comparative studies of breeding systems. Some practical implications for the plant breeder are discussed.

Seed plants exhibit an astonishing variety of reproductive systems, often within closely related groups. This diversity provides excellent material for evolutionary biologists interested in testing ideas about the evolution of different modes of reproduction, first exploited by Charles Darwin (1876, 1877), who contributed many of the original observations on which classifications of breeding systems are now based. The recent spate of research papers and reviews testifies to the continuing importance of this subject within evolutionary biology. Modes of reproduction are, of course, also of considerable importance in relation to the artificial improvement of economically important plant species, since the effectiveness and ease of genetic manipulations depend critically on factors such as whether individuals are cosexual or unisexual, or (in the case of cosexuals) self-compatible or incompatible. The consequences rather than the causes of the breeding system of a species are obviously of most significance for the practical breeder, but an understanding of its causes may help plant breeders to alter the breeding system to their own advantage.

This paper is concerned exclusively with the nature of the population processes involved in the evolution of reproductive systems in seed plants. In particular, I wish to examine the constraints imposed on these processes by the rules of genetic transmission. I shall argue that many features of breeding systems can be explained by such constraints. I shall not attempt a systematic survey of different types of systems, but will illustrate the general theme with some specific cases. In order to avoid confusion, I should point out that the existence of more severe constraints on the evolution of one type of reproductive system than on another does not necessarily mean that the first can never evolve: it simply means that its evolution requires more stringent conditions, so that it is less likely to arise in a group which has previously lacked it. Tests of evolutionary hypotheses based on comparative data are thus tests of the relative frequencies of different classes of evolutionary event, rather than of all-or-nothing effects (c.f. Charlesworth, 1984b). Some recent controversies, such as those concerning the respective roles of sexual selection, resource allocation and inbreeding avoidance (Thomson and Barrett, 1981; Givnish, 1982; Willson, 1979, 1982; Bawa, 1982; Lloyd, 1982) could perhaps have been avoided if this had been more clearly recognised.

GYNODIOECY AND ANDRODIOECY

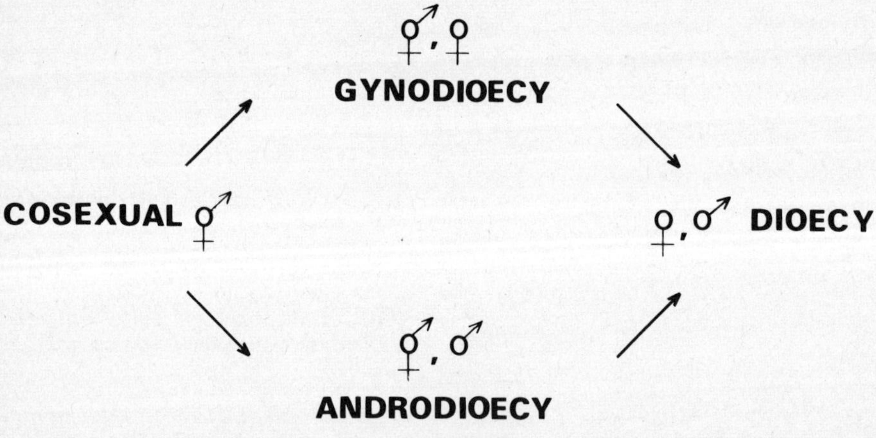

Figure 1

Evolution of gynodioecy, androdioecy and dioecy from cosexuality via male- and female-sterility mutations.

Gynodioecy and androdioecy are breeding systems in which populations contain a mixture of cosexual (hermaphrodite or monoecious) individuals, and females or males, respectively (Darwin, 1977). The simplest evolutionary path to gynodioecy is through the invasion of an initially entirely cosexual population by a male-sterility mutation; androdioecy could arise as a result of invasion by a female-sterility mutation (Fig. 1). The theoretical conditions for the successful establishment of these classes of mutation have been extensively studied. If it is assumed that the genetic factors concerned affect fitness solely through their effects on the sexual phenotype of their carriers, then simple discrete-generation population genetics analyses (Lewis, 1941; Lloyd, 1974, 1975; B. Charlesworth and D. Charlesworth, 1978) suggest that the factors listed in Table 1 are important in determining the success of the relevant mutations. It should be noted that δ incorporates the relative chances of successful fertilisation of selfed versus outcrossed ovules, as well as the adverse effects of inbreeding on progeny fitness. Also, the parameters $\underline{k}$ and $\underline{K}$ measure Darwin's (1877) "compensation" effect i.e. the effect of an automatic reallocation of resources to survival, growth on opposite sex function as a result of the loss of one sex function (Charnov et al., 1976).

Table 1. Breeding system parameters relevant to the evolutionary models.

Parameter	Definition
$\underline{s}$	Proportion of ovules of a cosexual plant that are self-fertilized.
δ	The difference between the probabilities that selfed and non-selfed ovules (which may or may not be fertilized) result in a viable and fertile progeny individual, as a proportion of the value for non-selfed ovules.
$\underline{k}$	The increase in female fertility of a male-sterile phenotype, as a proportion of the value for cosexuals.
$\underline{K}$	The increase in male fertility of a female-sterile phenotype, as a proportion of the value for cosexuals.

Both inbreeding avoidance and resource reallocation may thus provide advantages to loss of one sex function (Lloyd, 1975; Charlesworth and Charlesworth, 1978). It is, of course, possible that mutations affecting sex function may have more or less arbitrary pleiotropic effects on fitness, such as heterozygote advantage. Models incorporating such effects have been intensively studied (Ho and Ross, 1974; Ross and Weir, 1976; Ross, 1982; Gregorius et al., 1982). Although such pleiotropic effects can have a considerable influence on the outcome of selection, they must presumably vary in an unpredictable way across species. It therefore seems reasonable to ignore them when making predictions about the likelihood of the evolution of different breeding systems in terms of the parameters in Table 1.

If purely phenotypic selection effects are assumed, then the conditions for establishment in large populations of nuclear genes causing male or female sterility are, within certain limits, independent of the mode of gene action (dominant or recessive) and even of the number of loci and their linkage relations (Lloyd, 1975, 1977; B. Charlesworth and D. Charlesworth, 1978; Charlesworth, 1981; Gregorius and Ross, 1981). Formulae for the relative Darwinian fitnesses of different phenotypes, in terms of net genetic contributions to the next generation via pollen and ovules, are given by these authors, and can be used to predict the course of evolutionary change in specific models. For invasion by a male-sterility mutation, we require

$$k > 1 - 2 s \delta \qquad (1)$$

It is obviously easiest for male-sterility to invade (i.e. the smallest value of $\underline{k}$ is required) if the cosexual plants are partially self-fertilizing and experience some degree of inbreeding depression ($\underline{s}$, $\delta > 0$). More than a doubling of female fitness ($\underline{k} > 1$) is required in a self-incompatible species; intuitively, this arises from the fact that nuclear genes are transmitted equally through male and female reproduction (Fisher, 1930). A smaller value of $\underline{k}$ is required if there is some selfing and inbreeding depression, because the female plants avoid self-fertilization. If $2\underline{s}\,\delta > 1$, a male-sterility mutation can invade purely because of inbreeding avoidance, but this condition is very stringent. The equilibrium frequency of females among zygotes cannot exceed one-half, and it is likely to be much lower in most cases (Lloyd 1974, 1975).

The corresponding condition for invasion by a female-sterility mutation is

$$K > (1 + s - 2s\delta)/(1-s) \qquad (2)$$

In a fully self-incompatible species, a doubling of male fitness is required; this condition is symmetric with that for male-sterility. However, self-fertilisation leads to more stringent conditions (higher K values), even with inbreeding depression. This is due to the fact that self-fertilisation renders fewer ovules accessible to cross-pollination, so that the return from increased investment in male function is reduced (Lloyd, 1975; D. Charlesworth and B. Charlesworth, 1978). A low frequency of male plants among zygotes is the rule, even with $s = 0$.

These simple genetic considerations suggest that gynodioecy should be commoner than androdioecy, except in groups where the cosexual individuals are self-incompatible. Furthermore, it is far easier for gynodioecy due to nuclear genes to evolve in self-compatible species. However, a direct comparison of conditions (1) and (2) is difficult because k and K are unknown variables. It is know that, for the reason given in the preceding paragraph, cosexual species with high values of s will be selected to invest relatively more resources in female reproduction than species with low s values (D. Charlesworth and B. Charlesworth 1978, 1981). There is good comparative data supporting this prediction (Cruden, 1977; Schoen, 1982). It could thus be argued that cosexual species with high s values will tend to invest little in male reproduction, so that the k value of a male-sterility mutation will be small compared with the K value for a female sterility mutation. Such an asymmetry might conceivably outweigh the effects of s on conditions (1) and (2).

Theoretical studies of this question have been made by Charlesworth and Charlesworth (1981) and Charlesworth (1984a). The starting population is assumed to be composed entirely of cosexual plants which have adjusted their reproductive resources to the evolutionarily stable strategy (ESS), as defined by Maynard Smith (1972, 1982) and first applied to this problem by Charnov et al. (1976). The ESS corresponds to a division of resources between male and female reproduction, such that all mutations causing small changes in allocation are selectively disadvantageous (Fig. 2). The conditions under which male and female sterility mutations can invade are then obtainable in terms of s, δ, and the curve relating fitness through male reproduction to fitness through female reproduction.

Charlesworth and Charlesworth (1981) assumed that there was a trade-off between male and female fertility. With a concave relation between the two (Fig. 2), a male-sterility mutation can invade an ESS cosexual population, given suitable values of s and δ, but a female-sterility mutation can never invade, despite the existence of a strong bias towards female reproduction with moderate to high s values.

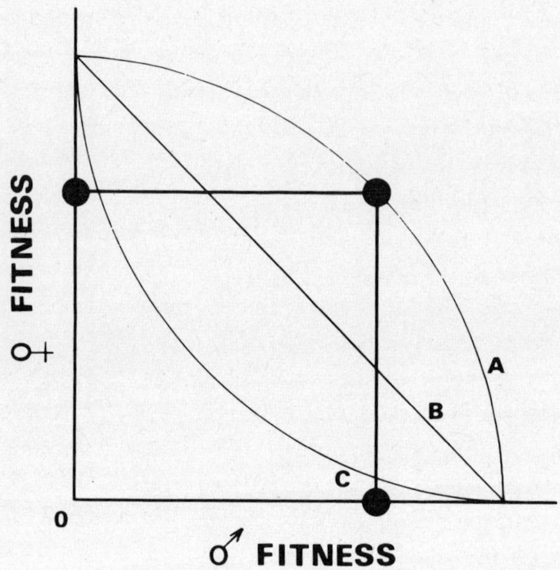

Figure 2

The trade-off between male and female fitnesses. A, B and C are concave, linear and convex curves, respectively. The circles indicate the location of a hypothetical ESS allocation of male and female function in a cosexual population.

An alternative model was studied by Charlesworth (1984a), who assumed that, in a long-lived perennial, there was a major influence of female reproductive activity in one season on growth and survival to the next season, whereas male reproduction had a negligible effect on these traits. This takes account of the fact that a major part of female investment in seed plants goes into ripening of the fruit (Lovett Doust and Lovett Doust, 1981) and so cannot be effectively traded against male function (Charnov et al., 1976), but is a heavy drain on resources that could be used for growth or maintenance (e.g. Meagher and Antonovics, 1982). This model predicts a rather low ESS allocation of resources to female reproduction unless $\underline{s}$ is very high, and so the large boost to fitness accruing from a reduction in female investment is insufficient to permit a female-sterility mutation to invade (Charlesworth, 1984a), unless there is a convex relationship between female reproduction and the other fitness components. (For reasons discussed by Charlesworth [1984a], such a relationship seems unlikely, unless the species concerned has recently experienced a change in ecology that affects its reproductive biology).

The above arguments assume nuclear inheritance of the genes concerned. Cytoplasmic inheritance of male-sterility is very favourable to the establishment of gynodioecy, since cytoplasmic factors are not transmitted through pollen and hence suffer no loss from male-sterility (Lewis, 1941). The condition corresponding to (1) for a cytoplasmic factor is simply

$$k + s\delta > 0 \qquad (3)$$

Conversely, cytoplasmic genes inducing female sterility are obviously doomed, which is another factor telling against the evolution of androdioecy (Lloyd, 1975; B. Charlesworth and D. Charlesworth, 1978). It might therefore be expected that cytoplasmic control of gynodioecy should be very common (Lewis, 1941), particularly as cytoplasmic genes causing male sterility are commonly used in agricultural species (Beale and Knowles, 1978), so that there is no difficulty in invoking such mutations. In fact, genetic studies of gynodioecious species usually reveal a rather messy pattern of inheritance, with evidence of both nuclear and cytoplasmic inheritance. There are few, if any, clear-cut examples of strict nuclear or cytoplasmic inheritance (Ross, 1978; Charlesworth, 1981; Van Damme, 1983). This has stimulated the development of models of joint nuclear and cytoplasmic inheritance, based on the well-studied cytoplasmic male sterility factors of maize and nuclear restorer genes, which cancel their effects (Laughnan and Gabay, 1978). These models indicate that mixed inheritance of male-sterility can generate a stable polymorphism (gynodioecy), provided that the restorer genes cause a drastic pleiotropic reduction in fitness (Charlesworth and Ganders, 1979; Charlesworth, 1981; Delannay et al., 1981). Such polymorphisms are possible even in self-incompatible species. There is evidence for such side-effects of restorer genes in maize (Laughnan and Gabay, 1978).

Despite these complications, the basic conclusion that gynodioecy should be commonest in species where the cosexual plants are self-compatible, and that androdioecy should be rare or non-existent, remain valid. They also agree remarkably well with the data. The overwhelming majority of examples of gynodioecy involve self-compatible cosexuals (B. Charlesworth and D. Charlesworth, 1978; Charlesworth, 1981). There are two well-studied exceptions: Plantago lanceolata (Ross, 1969, 1973; Van Damme, 1983; Van Damme and Van Delden, 1982), and Hirschfeldia incana (Horovitz and Beiles, 1980). Both of these involve mixed nuclear and cytoplasmic inheritance. Androdioecy does appear to be extremely rare, although a number of cases have been reported in the literature (Ross, 1982; Charlesworth, 1984a). However, all cases that have been thoroughly investigated seem to involve cryptic dioecy, in which the apparently hermaphrodite individuals are functionally male, frequently producing inaperturate, non-functional pollen in approximately the same quantities as in the females (Charlesworth, 1984a). The species concerned use pollen rather than nectar as a pollinator reward, so that it seems likely that mutations completely abolishing male function would be disadvantageous. It is unclear whether any genuine examples of androdioecy, in the sense of a genetic polymorphism for female sterility, exist.

DIOECY

Dioecy, the co-occurrence of unisexual male and female plants in the same population, is a common breeding system in Gymnosperms (Givnish, 1980), and occurs sporadically in many Angiosperm families, although a few are exclusively dioecious (Lewis, 1942; Charlesworth, 1984b). The comparative evidence suggests that dioecy usually evolves from a cosexual, homomorphic ancestral state (Darwin, 1877; Baker, 1959a; Ornduff, 1966; Lloyd, 1979; Beach and Bawa, 1980).

I will first consider the path by which dioecy evolves from a homomorphic ancestor, assuming that the final produce involves genetic rather than environmental sex-determination. As shown in Fig. 1, the simplest path involves two mutations, one causing male-sterility and the other female-sterility. In reality, several mutations with less drastic effects on sex function may be involved (D. Charlesworth and B. Charlesworth, 1978; Ross, 1982), but the main principles involved are brought out clearly by this simple case. From the arguments presented above, it seems most likely that the initial step in the evolution of dioecy would involve a male-sterility rather than a female-sterility mutation, so that gynodioecy rather than androdioecy would be the intermediate stage on the path to dioecy. Conversion of gynodioecy into dioecy involves the spread of a female-sterility mutation (Fig. 1).

It is clear that gynodioecy involving strict cytoplasmic inheritance of male sterility cannot evolve into dioecy, since females could not segregate male and female progeny. Either joint nuclear-cytoplasmic or strict nuclear inheritance of male-sterility are required. Numerical studies by Delannay et al. (1981) and Charlesworth (1981) have not yielded conditions under which nuclear-cytoplasmic inheritance of gynodioecy can evolve into true dioecy, although (under stringent conditions) a female-sterility mutation can invade and come to an equilibrium with males, females and cosexuals present. As suggested by Ross (1978), the prevalence of gynodioecy with joint nuclear-cytoplasmic inheritance probably reflects its stability to invasion by female-sterility mutations.

Nuclear inheritance of gynodioecy can, in contrast, lead to the evolution of full dioecy relatively easily. If females are present in the equilibrium population at a significant frequency, then investment in male function by other members of the population becomes more profitable than in a purely cosexual population. This follows from Fisher's (1930) principle that the average fitness through male and female reproduction is the same with nuclear gene inheritance. Using the model of Table 1, the condition for invasion of an equilibrium, gynodioecious population by a female-sterility mutation that converts cosexuals into males, but has no effect on females, is (B. Charlesworth and D. Charlesworth, 1978) as follows

$$K > \frac{1 + s - 2s\delta}{k + 2s\delta - s} \qquad (4)$$

Inspection of this condition shows that it is always easier to satisfy than condition (2) for invasion of a cosexual population by a female-sterility mutation, and that high values of $\underline{k}$ and δ (implying a high frequency of females in the gynodioecious population) are most favourable to invasion. Condition (4) also implies that $\underline{K} > 0$ is required for invasion, i.e. gynodioecy cannot evolve into dioecy unless female-sterility mutations generate an increase in male fitness. Calculations of the values of K and k for the case when the cosexuals are at an ESS for resource allocation show that condition (4) is most easily satisfied if the curve relating female and male fitnesses is not too concave, and inbreeding depression, as measured by δ, is high (Charlesworth and Charlesworth, 1981).

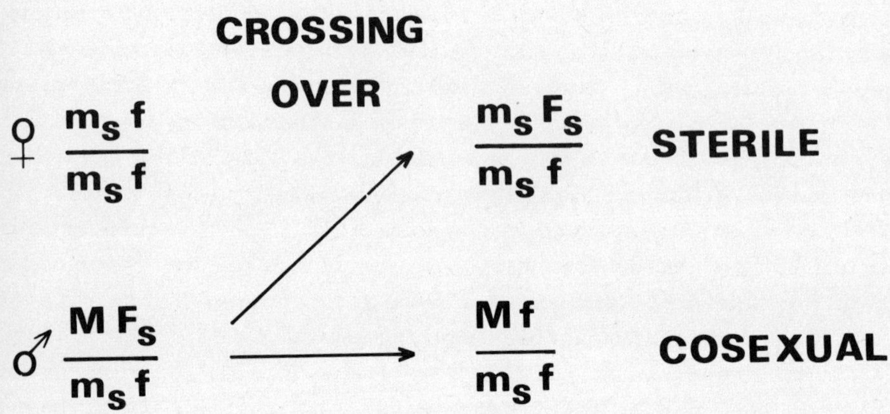

Figure 3

The genetic constitution of a nearly completely dioecious population, with low frequencies of cosexual and sterile plants produced by crossing over. M and M_s are dominant male-fertility and (recessive) male-sterility alleles; F_s and f are (dominant) female-sterility and (recessive) female-fertility alleles.

If the female-sterility mutation is expressed in females as well as in cosexuals, as seems likely to be the commonest situation, then sterile individuals will be produced unless linkage between the two loci is complete. If the initial male-sterility mutation is recessive, this leads to a "linkage constraint" (B. Charlesworth and D. Charlesworth, 1978), such that a female-sterility mutation can invade only if it is sufficiently tightly linked to the male-sterility locus. The evolution of full dioecy in this case is limited to species in which mutations to male- and female-sterility can occur at closely linked loci (recombination frequencies < 5% or so). Recombination frequencies that are only just below the critical value for invasion yield equilibria in which cosexual and sterile plants are produced by recombination in significant frequencies (Fig. 3). As shown by B. Charlesworth and D. Charlesworth (1978), in such cases there is selection for chromosome rearrangements or other devices that restrict crossing over between the loci. If the female-sterility mutation is fully dominant, this leads eventually to the evolution of full dioecy with male heterogamety (Fig. 3). This gives the appearance of a single locus determining sex, but in reality at least two loci are involved. If the female-sterility mutation is only partially dominant, there will be selection for modifiers that render it dominant, since cosexuals are at a selective disadvantage (this process has never been formally modelled, however). Again, the end-product is full dioecy. If chromosomal rearrangements or chiasma localisation are involved in the suppression of crossing over between the "X" and "Y" chromosomes, recombination between them may be suppressed to a considerable extent (Westergaard, 1958). This may lead ultimately to the "degeneration" of much of the Y chromosome and to dosage compensation for sex-linked genes (Charlesworth, 1978; Bull, 1983). Heteromorphic sex chromosomes occur in some groups of plants (Westergaard, 1958).

The above pattern of male heterogamety with an essentially two-locus determination of sex appears to be widespread in Angiosperms (Westergaard, 1958). B. Charlesworth and D. Charlesworth (1978) have argued that a route involving recessive male-sterility is more likely than that involving dominant male-sterility, since recessive male-sterility mutations are known to be much more frequent than dominant ones. Female heterogamety can, however, evolve if the initial male-sterility mutation is dominant and the female-sterility mutation is recessive. Male heterogamety is much commoner than female heterogamety among dioecious species of flowering plants (Westergaard, 1958; B. Charlesworth and D. Charlesworth, 1978; Bull, 1983), which fits the prediction of this model. If the dominance relations of the male- and female-sterility mutations are not complementary, however, full dioecy cannot evolve and the population will stall at a subdioecious stage in which a mixture of males, females and cosexuals are present (Ross and Weir, 1976; B. Charlesworth and D. Charlesworth,

1978). Not surprisingly, this kind of situation is quite common (Darwin, 1877; Ross, 1978). Another factor causing a tendency to evolve sub-dioecy arises from the fact that genes of minor effect on female fertility may well be selected, rather than a full female-sterility mutation (B. Charlesworth and D. Charlesworth, 1978). Unless genes of the right linkage and dominance relationships are incorporated into the population, a highly variable sexual phenotype among the "males" might thus be expected, and is in fact commonly observed (Darwin, 1877; Westergaard, 1958). This difference between male and female variability applies to species which were originally hermaphrodite and self-compatible, where a mutation of large effect on male fertility is at a selective premium over genes with small effects, since the latter may not reduce selfing rates by much. With monoecy as a starting-point, however, genes which modify the proportion of male versus female flowers in the direction of more femaleness are likely to have a roughly proportionate effect on selfing, and a gradual evolution of femaleness is possible (D. Charlesworth and B. Charlesworth, 1978). In this case, therefore, variable expression is expected in both males and females, and seems frequently to be observed (Lloyd 1973, 1980b).

Dioecy may also evolve from a distylous starting-point (Darwin, 1877; Baker, 1959a; Lloyd, 1979; Ornduff, 1966; Beach and Bawa, 1980). The theory of this process has been formalised by Lloyd (1979) and by Casper and Charnov (1982), based on a suggestion by Baker (1959a). Consider a population segregating for long-styled and short-styled morphs, which are self-incompatible but mutually cross-compatible (pins and thrums). If one of the forms (e.g. thrum) has a slightly lower ovule fertility for some reason (such as lack of accessibility of its stigma to pollinators [Baker, 1959a]), then a mutant causing the opposite form to increase its investment in female reproduction at the expense of male production will be favoured. Conversely, mutants causing an increase in male reproduction in the other morph will be favoured. As before, the mutations concerned would have to be epistatic or tightly linked to the distyly locus. Clearly, this process involves resource re-allocation rather than inbreeding depression, but Ornduff (1966) and Lloyd (1979) have suggested that evolution towards dioecy could also occur if one of the distyly morphs were more self-compatible than the other, in which case evolution towards increased femaleness would be favoured, triggering evolution towards maleness of the alternative form. There is evidence for such a process in Nymphoides.

Except for the pathway from heterostyly, the theoretical arguments presented above suggest that dioecy is most likely to evolve from a self-compatible, partially self-fertilising cosexual ancestor, although evolution from a self-incompatible ancestor is not impossible. The question of whether or not the data bear out this expectation has been vigorously debated in recent years, with some authors emphasizing resource reallocation and proposing that inbreeding avoidance plays a minimal role in the evolution of dioecy (Willson, 1979, 1982; Givnish, 1980, 1982). As discussed above, both inbreeding avoidance and resource reallocation almost certainly play a role in most instances; it is clear that dioecy cannot evolve in the absence of resource reallocation, but that selfing and inbreeding depression greatly facilitate its evolution. One would, therefore, expect an association between a self-compatible ancestral state and the evolution of dioecy. Baker (1959) and Thomson and Barrett (1981) have presented some comparative data supporting this conclusion, but Givnish (1982) has reported that his extensive survey of Angiosperm breeding systems shows that there is no greater frequency of families with both dioecious and self-compatible cosexual members than would be expected by chance, knowing the net frequencies of families with dioecy and self-compatibility. His testing procedure has been criticised by Charlesworth (1984b), who pointed out that the evidence concerning the distribution of self-incompatibility is very scanty. On the basis of a more detailed survey of the literature, she concluded that there was indeed no significant tendency for families with dioecy to lack self-incompatible species. On the other hand, the frequency of dioecious genera is much lower (7%) among families known to have some homomorphic self-incompatible species, compared with a value of 13% for families where only self-compatibility is known. This difference is highly significant statistically. The frequency of dioecious genera among families with some distylous species is also high (13%). Furthermore, there is good evidence that cosexuals in subdioecious species, and in the close relatives of dioecious species, are nearly always self-compatible. These observations suggest that it is easier to evolve dioecy from a self-compatible or a distylous ancestral state than from a homomorphic, self-compatible state, as predicted by the models.

SELF-FERTILISATION AND ASEXUALITY

The evolution of inbreeding versus outbreeding and of asexual (parthenogenetic) reproduction can only be understood in relation to the concept of the "cost of sex" (Maynard Smith, 1971), often referred to as the "cost of meiosis" (Williams, 1975). Consider, for example, a cosexual species with 100% outcrossing. A dominant gene which causes its carriers to produce seed exclusively by self-fertilisation, but which does not affect their seed fertility or contribution to the pollen pool used in outcrossing by other plants, will increase its frequency from p to (3/2)p each generation when p is small, provided that inbreeding depression is absent (Fisher, 1941). Similarly, a dominant mutation causing the asexual production of seed whose genotype is identical to that of the mother plant will increase at the same rate, provided that it contributes normally to the pollen pool and has normal fertility through seed (Jaenike and Selander, 1979; Charlesworth, 1980; Lloyd, 1980a). The advantage of both these classes of mutation is that their seeds transmit purely maternal genes, instead of one maternal and one paternal gene, and hence have a doubled representation via female reproduction.

These types of calculation can be extended to more complex situations (Charlesworth, 1980; Lloyd, 1980; Uyenoyama, 1984). For example, it can be shown that a gene causing an increased level of self-fertilisation or asexual reproduction will spread in a population with $\underline{s} < 1$ (if inbreeding depression is absent), provided its carriers make some contribution to the outcrossing pollen pool. Its selective advantage, as measured by the initial rate of increase in gene frequency, declines with the population value of $\underline{s}$, and with the proportion of a plant's pollen expended in self-fertilisation. The advantage is zero if the population is wholly self-fertilising, or if all of a plant's pollen is used up in selfing (Nagylaki, 1976; Charlesworth, 1980; Lloyd, 1980a). In a dioecious species with a 1:1 sex ratio, a mutation causing females to produce exclusively female offspring experiences an advantage such that its frequency, when rare, doubles each generation (Maynard Smith, 1971). These results suggest that asexual reproduction should evolve most frequently in dioecious or self-incompatible species (Charlesworth, 1980; Lloyd, 1980a), a prediction that appears to be confirmed by the comparative data of Gustafsson (1947).

Of course, the advantage of both asexual reproduction and self-fertilisation can be modified by other factors, notably the relative reproductive success of selfed or asexual ovules compared with normally produced ones. A low efficiency of pollination will tend to increase the net reproductive success of selfed or asexual ovules relative to outcrossed ones, so that we would expect a tendency for the evolution of self-fertilisation and asexual reproduction to be correlated with habitats in which the chance of cross-fertilisation is low. Such a correlation has been noted many times (e.g. Henslow, 1879; Baker, 1959b, 1967; Levin, 1975; Glesener and Tilman, 1978; Lloyd, 1980a,c; Bell, 1982), although it has been variously interpreted.

As far as self-fertilisation is concerned, the parameter in Table 1 combines the antagonistic effects of inbreeding depression and assurance of reproduction by selfing into a single measure of the relative success of selfed ovules. In the kind of model described above, genes causing an increased rate of selfing will be favoured when $\delta < 1/2$, and eliminated when $\delta > 1/2$ (Charlesworth, 1980; Lloyd, 1980a). This model is probably most appropriate for genes that alter the selfing rate by changing the distance between stigmas and anthers (Breese, 1959), or for genes that cause protandry or protogyny. Variants at the locus or loci controlling an incompatibility locus may have very different dynamics. For example, D. Charlesworth and B. Charlesworth (1979b) showed that a population at equilibrium for a homomorphic, gametophytic self-incompatibility system, controlled by a single S-locus, can be invaded by an S_f allele that is compatible with all alleles (including itself) in both pollen and style, when < 2/3. The stability of the S-allele system against such invasion increases with the number of S-alleles present; a substantially higher value of δ permits invasion if the number of alleles is of the order of 10. Similar results are obtained for mutations that cause loss of the reaction of one of the S-alleles in pollen only or style only (equivalent to the pollen-part mutations of Lewis [1951] and style reaction mutations of Pandey [1956]). Conversely, the conditions for invasion of a self-compatible population by an S-allele requires values much greater than 1/2. It is unlikely that S_f or other self-compatibility alleles will be entirely eliminated unless the number of alleles is very high; this is consistent with reports of naturally occurring self-compatibility variants at the S-locus in several species (D. Charlesworth and B. Charlesworth, 1979b). On the other hand, modifiers at other loci which abolish the incompatibility reactions of the S-alleles are able to invade only if $\delta < 1/2$. (D. Charlesworth and B. Charlesworth, 1979b).

The need to consider the genetics of individual systems when theorising about the evolution of inbreeding versus outbreeding is brought out further when heteromorphic self-incompatibility systems (distyly and tristyly) are compared with each other, and with homomorphic systems. In distylic systems, there are two "alleles" at a supergene controlling anther position, pollen reaction, stigma position, and stigma reaction. The incompatibility system is sporophytic and most easily interpreted in terms of two alleles at a locus controlling pollen reaction (P and p), and two alleles at a stigma reaction locus (G and g), such that pollen from gp/gp plants is compatible with the stigmas of GP/gp plants, and vice-versa. The tight linkage between the loci enables the maintenance of just two incompatibility types in most populations, although occasional crossovers result in the production of self-compatible recombinant types (Gp and gP). Such types can invade a distylous population if $\delta < 1/2$, but otherwise are kept at low frequencies by selection (B. Charlesworth and D. Charlesworth, 1979). The breakdown of heterostyly due to invasion by these self-compatible morphs shows an interesting asymmetry with respect to dominance; rare gP - carrying plants will fertilise gp/gp stigmas, and gP will be expressed in every progeny individual that received it, whereas Gp/gp pollen can fertilise only GP/gp stigmas, in which case only one-half the progeny that receive Gp will express it. Hence, gP spreads more rapidly than Gp when $\delta < 1/2$. The comparative data on secondarily self-compatible relatives of distylous species shows that the recombinant with the dominant pollen type is most frequently established (B. Charlesworth and D. Charlesworth, 1979). Recent studies of the self-compatible, locally common, long homostyle variant of the primrose, *Primula vulgaris*, have shown that it is highly selfing ($s \simeq 0.90$) and (in some years at least) has a higher seed set than the self-incompatible morphs in the same populations. This suggests that its spread is due to the advantage of assurance of pollination overcoming the effects of inbreeding depression (Piper et al., 1984).

In tristylous populations, there are three mutually compatible forms, each with two sets of anthers at different heights and differing with respect to stigma and anther position as well as in incompatibility reaction (Darwin, 1877). The genetics of tristyly seems to be similar in most species that have been investigated, involving two epistatically interacting loci controlling anther and stigma positions. The incompatibility reactions of pollen and stigma seem to be directly controlled by anther and stigma position (Charlesworth, 1979). Breakdown of tristyly can occur by several routes, either involving loss of an allele at one of the loci controlling morphology, or by the selection of modifiers that alter morphology or the incompatibility reaction (Charlesworth, 1979; Heuch, 1980). However, selection for breakdown always seems to require $\delta < 1/2$.

The evolution of heterostyly is a much more difficult problem than its loss. The model of D. Charlesworth and B. Charlesworth (1979a) for distyly assumes that the ancestral condition is a homomorphic, self-compatible state (e.g., Gp). The evolution of incompatibility involves the selection, first, of a mutation changing the pollen reaction ($\underline{p} \rightarrow \underline{P}$) and, second, of a mutation changing the stigma reaction ($\underline{G} \rightarrow \underline{g}$), followed by selection of mutations altering morphology. The first step is equivalent to invasion by a male-sterility mutation, but without any resource reallocation advantage, and requires the very stringent condition $\underline{s}\delta > 1/2$. Muenchow (1981, 1982), however, has argued that the incompatibility system of distyly is similar to that in homomorphic, sporophytic incompatibility systems, and proposed that it evolved by the loss of alleles from such a system. Her theory has been criticised by Charlesworth (1982). Charlesworth (1979) constructed a model involving three loci for the evolution of tristyly, two of which are the same as those presently involved in the heteromorphism, and one of which is assumed to have experienced an allelic substitution and hence is not detectable today. Depending on the exact pathway assumed, tristyly can evolve either if $\delta > 1/2$ or $\underline{s}\delta > 1/2$; presumably the path involving the lighter of these condition is more likely. It is extremely hard to see how these theories can be tested, except by examining the validity of their genetical assumptions.

The assumption that there is a fixed parameter, δ, which measures the reproductive success of selfed ovules is, of course, unrealistic in view of the fact that inbreeding depression is a complex phenomenon involving contributions from many loci (Falconer, 1981). Maynard Smith (1977) and Lande and Schemske (1985) have attempted to construct more realistic models. Nevertheless, the main conclusions derived above seem to remain approximately valid. In the case of inbreeding depression due to homozygosity for recessive or partially recessive deleterious alleles maintained by mutation pressure, the degree of inbreeding depression in a partially selfing population declines with the value of s (Charlesworth, unpub; Lande and Schemske, 1984). Hence, the spread of a gene causing an altered selfing rate will tend to be somewhat self-accelerating, since δ will increase or decrease according to whether s is decreased or increased. This leads to the prediction that selection will usually favour extreme over intermediate values of s, although some situations in which an intermediate value of s is stable have been proposed (Maynard Smith, 1977; Lloyd, 1980b). The available comparative data on natural rates of selfing seem to bear out this prediction (Schemske and Lande, 1985). It would be extremely valuable to have more information on natural selfing rates, such as may be easily obtained with electrophoretic techniques (Brown, 1979, this volume).

CONCLUSIONS

The examples described above demonstrate that the details of the genetic system play a major role in determining the paths by which plant reproductive systems evolve, and also affect the nature of the final phenotypes observed. Arguments based purely on phenotypic optimisation models or on game theory (Charnov, 1982; Maynard Smith, 1982), although extremely useful in many instances, ignore these constraints, and hence do not always do justice to the complexities of natural systems. Although there are many areas of uncertainty and controversy, the application of population genetics models to these questions in recent years has, in my view, considerably clarified our ideas about the evolution of plant reproductive systems and demonstrated the inadequacy of the older "advantage-to-the-species" concepts popularised by Darlington (1939) and Stebbins (1950). In some cases, the predictions of the models seem to be verified remarkably well by the genetic and comparative data: e.g., the rarity of androdioecy, the nature of genetic sex-determination in dioecious species, and the mode of breakdown of distyly. It is to be hoped that further genetic and ecological studies will provide data for more rigorous tests, and will stimulate the development of better theories.

The practical plant breeder will naturally wish to be convinced of the relevance of these rather academic questions to his problems. A couple of examples may be persuasive. First, the models of the evolution of dioecy suggest that it should often be possible to produce self-fertile cosexual plants which breed true, as a result of the appropriate crossovers between the sex-determining loci (Fig. 3). Several cases of this have in fact occurred in cultivated plants (Lewis, 1942; Westergaard, 1958). The use of irradiation to induce chromosome rearrangements, or other types of genetic exchange, between regions of restricted crossing over might be seriously considered in cases where it is wished to develop self-fertile strains, and where direct selection has proved ineffective. Second, the existence of cryptic dioecy, in which functional males are morphologically identical to hermaphrodites, could well pose problems for the breeder if not diagnosed correctly. Several examples of this are known in tropical trees (Charlesworth, 1984a) where the occurrence of an approximately 1:1 ratio of hermaphrodite to female plants provides strong *prima facie* evidence for functional dioecy in these cases. Finally, the theory of *S*-allele systems suggests that self-compatibility variants may often be stably maintained in species which are predominantly self-compatible. Screening natural material for such variants might well be a profitable strategy for developing self-fertile strains.

ACKNOWLEDGEMENTS

I thank Deborah Charlesworth and David Lloyd, who contributed many of the ideas presented here, for their comments on the manuscript

REFERENCES

Baker, H.G. 1959a. Reproductive methods as factors in speciation in flowering plants. *Cold Spring Harb. Symp. Quant. Biol.* 24: 177-190.

Baker, H.G. 1959b. The contribution of autecological and genecological studies to our knowledge of the past migrations of plants. *Amer. Nat.* 93: 255-272.

Baker, H.G. 1967. Support for Baker's Law - as a rule. *Evolution* 20: 349-368.

Bawa, K.S. 1982. Outcrossing and the incidence of dioecism in island floras. Amer. Nat. 119: 866-871.

Beach, J.H. and K.S. Bawa. 1980. Role of pollinators in the evolution of dioecy from distyly. Evolution 34: 1138-1143.

Beale, G. and J. Knowles. 1978. Extranuclear Genetics. Edward Arnold, London.

Bell, G. 1982. The Masterpiece of Nature. Croom Helm, London.

Breese, E.L. 1959. Selection for differing degrees of out-breeding in Nicotiana rustica. Ann. Bot. N.S. 23: 331-344.

Brown, A.H.D. 1979. Enzyme polymorphism in plant populations. Theoret. Pop. Biol. 15: 1-42.

Bull, J.J. 1983. Evolution of Sex Determining Mechanisms. Benjamin Cummings, Menlo Park.

Casper, B.B. and E.L. Charnov. 1982. Sex allocation in heterostylous plants. J. Theor. Biol. 96: 143-149.

Charlesworth, B. 1978. Model for evolution of Y chromosomes and dosage compensation. Proc. Natl. Acad. Sci. USA 75: 5618-5622.

Charlesworth, B. 1980. The cost of sex in relation to mating system. J. Theor. Biol. 84: 655-671.

Charlesworth, B. and D. Charlesworth. 1978. A model for the evolution of dioecy and gynodioecy. Amer. Nat. 112: 975-997.

Charlesworth, B. and D. Charlesworth. 1979. The maintenance and breakdown of distyly. Amer. Nat. 114: 499-513.

Charlesworth, D. 1979. The evolution and breakdown of tristyly. Evolution 33: 486-498.

Charlesworth, D. 1981. A further study of the problem of the maintenance of females in gynodioecious species. Heredity 46: 27-39.

Charlesworth, D. 1982. On the nature of the self-incompatibility locus in homomorphic and heteromorphic systems. Amer. Nat. 119: 732-735.

Charlesworth, D. 1984a. Androdioecy and the evolution of dioecy. Biol. J. Linn. Soc. 23: 333-348.

Charlesworth, D. 1984b. Distribution of dioecy and self-incompatibility in Angiosperms. In Evolution - Essays in Honour of John Maynard Smith (P.J. Greenwood and M. Slatkin, eds.) Cambridge University Press, Cambridge.

Charlesworth, D. and B. Charlesworth. 1978. Population genetics of partial male-sterility and the evolution of monoecy and dioecy. *Heredity* 41: 137-153.

Charlesworth, D. and B. Charlesworth. 1979a. A model for the evolution of distyly. *Amer. Nat.* 114: 467-498.

Charlesworth, D. and B. Charlesworth. 1979b. The evolution and breakdown of S-allele systems. *Heredity* 43: 41-55.

Charlesworth, D. and B. Charlesworth. 1981. Allocation of resources to male and female functions in hermaphrodites. *Biol. J. Linn. Soc.* 15: 57-74.

Charlesworth, D. and F.R. Ganders. 1979. The population genetics of gynodioecy with cytoplasmic-genic male-sterility. *Heredity* 43: 213-218.

Charnov, E.L. 1982. *The Theory of Sex Allocation*. Princeton University Press, Princeton.

Charnov, E.L., J. Maynard Smith and J.J. Bull. 1976. Why be a hermaphrodite? *Nature* 263: 125-126.

Cruden, R.W. 1977. Pollen-ovule ratios: a conservative indicator of breeding systems in flowering plants. *Evolution* 31: 32-46.

Darlington, C.D. 1939. *Evolution of Genetic Systems*. Cambridge University Press, Cambridge.

Darwin, C.R. 1876. *The Effects of Cross and Self Fertilisation in the Vegetable Kingdom*. John Murray, London.

Darwin, C.R. 1877. *The Different Forms of Flowers on Plants of the Same Species*. John Murray, London.

Delannay, X., P.H. Gouyon and G. Valdeyron. 1981. Mathematical study of gynodioecy with cytoplasmic inheritance under the effect of a nuclear restorer gene. *Genetics* 99: 169-181.

Falconer, D.S. 1981. *Introduction to Quantitative Genetics* (2nd edn.) Longman, London.

Fisher, R.A. 1930. *The Genetical Theory of Natural Selection*. Oxford University Press, Oxford.

Fisher, R.A. 1941. Average excess and average effect of a gene substitution. *Ann. Eugen.* 11: 53-63.

Givnish, T.J. 1980. Ecological constraints on the evolution of breeding systems in seed plants: dioecy and dispersal in gymnosperms.
Evolution 34: 959-972.

Givnish, T.J. 1982. Outcrossing versus ecological constraints in the evolution of dioecy.
Amer. Natur. 119: 849-865.

Glesener, R.R. and D. Tilman. 1978. Sexuality and the components of environmental uncertainty: Clues from geographic parthenogenesis in terrestrial animals.
Amer. Nat. 112: 659-673.

Gregorius, H-R. and M.D. Ross. 1981. Selection in populations of effectively infinite size: I. Realised genotypic fitnesses.
Math. Biosci. 54: 291-307.

Gustafsson, A. 1947. Apomixis in higher plants. Part II. The causal aspect of apomixis.
Lunds Univ. Arsskrift 43: 166-178.

Henslow, G. 1879. On the self-fertilisation of plants.
Trans. Linn. Soc. Lond . Ser. 2. Bot. 1: 317-398.

Heuch, I. 1980. Loss of incompatibility types in finite populations of the heterostylous plant Lythrum salicaria.
Hereditas 92: 53-57.

Ho, T-Y. and M. D. Ross. 1974. Maintenance of males and females in hermaphrodite populations.
Heredity 32: 113-118.

Horovitz, A. and A. Beiles. 1980. Gynodioecy as a populational strategy for increasing reproductive output.
Theor. Appl. Genet. 57: 11-15.

Jaenike, J.R. and R.K. Selander. 1979. Evolution and ecology of parthenogenesis in earthworms.
Amer. Zool. 19: 727-737.

Lande, R. and D.W. Schemske. 1985. The evolution of self-fertilisation and inbreeding depression in plants. I. Genetic models.
Evolution 39: 24-40.

Laughnan, J.R. and S.J. Gabay. 1978. Nuclear and cytoplasmic mutations to fertility in S male-sterile maize. In Maize Breeding and Genetics. (D.B. Walden, ed.)
John Wiley, New York.

Levin, D.A. 1975. Pest pressure and recombination systems in plants.
Amer. Nat. 109: 437-451.

Lewis, D. 1941. Male sterility in populations of hermaphrodite plants.
New Phytol. 40: 56-63.

Lewis, D. 1942. The evolution of sex in flowering plants.
Biol. revs. 17: 46-67.

Lewis, D. 1951. Structure of the incompatibility gene. III. Types of spontaneous and induced mutation.
Heredity 5: 399-414.

Lloyd, D.G. 1973. Sexual dimorphism in Cotula.
Genetics 74: 161.

Lloyd, D.G. 1974. Theoretical sex ratios of dioecious and gynodioecious angiosperms.
Heredity 32: 11-34.

Lloyd, D.G. 1975. The maintenance of gynodioecy and androdioecy in angiosperms.
Genetica 45: 325-339.

Lloyd, D.G. 1977. Genetic and phenotypic models of natural selection.
J. Theor. Biol. 69: 543-560.

Lloyd, D.G. 1979. Evolution towards dioecy in heterostylous populations.
Plant Syst. Evol. 131: 71-80.

Lloyd, D.G. 1980a. Benefits and handicaps of sexual reproduction.
Evol. Biol. 15: 69-111.

Lloyd, D.G. 1980b. The distributions of gender in four Angiosperm species illustrating two evolutionary pathways to dioecy.
Evolution 34: 123-134.

Lloyd, D.G. 1980c. Demographic factors and mating patterns in plant populations. In Demography and Evolution in Plant Populations (O.T. Solbrig, ed.).
Blackwell, Oxford.

Lloyd, D.G. 1982. Selection of combined versus separate sexes in seed plants.
Amer. Nat. 120: 571-585.

Lovett Doust, J. and L. Lovett Doust. 1982. Parental strategy: gender and maternity in higher plants.
Biosciences 33: 180-186.

Maynard Smith, J. 1971. The origin and maintenance of sex. In Group Selection (G.C. Williams, ed.).
Aldine Atherton, Chicago.

Maynard Smith, J. 1972. On Evolution.
Edinburgh University Press, Edinburgh.

Maynard Smith, J. 1982. Evolution and the Theory of Games.
Cambridge University Press, Cambridge.

Meagher, T.R. and J. Antonovics. 1982. The population biology of Chamaelirium luteum, a dioecious member of the lily family. III. Life history studies.
Ecology 63: 1690-1700.

Muenchow, G. 1981. An S-locus model for the distyly supergene.
Amer. Nat. 118: 756-760.

Muenchow, G. 1982. A loss-of-alleles model for the evolution of distyly.
Heredity 49: 81-94.

Nagylaki, T. 1976. A model for the evolution of self-fertilization and vegetative reproduction.
J. Theor. Biol. 58: 55-58.

Ornduff, R. 1966. The origin of dioecism from heterostyly in Nymphoides (Menyanthaeceae).
Evolution 20: 309-314.

Pandey, K.K. 1956. Mutations of self-incompatibility alleles in Trifolium pratense and T. repens.
Genetics 41: 327-343.

Piper, J.G., B. Charlesworth and D. Charlesworth. 1984. A high rate of self-fertilization and increased seed fertility of homostyle primroses.
Nature 310: 50-51.

Ross, M.D. 1969. Digenic inheritance of male sterility in Plantago lanceolata.
Can. J. Genet. Cytol. 11: 739-744.

Ross, M.D. 1970. Evolution of dioecy from gynodioecy.
Evolution 24: 827-828.

Ross, M.D. 1973. Inheritance of self-incompatibility in Plantago lanceolata.
Heredity 30: 169-176.

Ross, M.D. 1978. The evolution of gynodioecy and subdioecy.
Evolution 32: 174-188.

Ross, M.D. 1982. Five evolutionary pathways to subdioecy.
Amer. Nat. 119: 297-318.

Ross, M.D. and B. Weir. 1976. Maintenance of males and females in hermaphrodite populations and the evolution of dioecy.
Evolution 30: 425-441.

Schemske, D.W. and R. Lande. 1985. The evolution of self-fertilization and inbreeding depression in plants. II. Empirical observations.
Evolution 39: 41-52.

Schoen, D.J. 1982. The breeding system of Gilia achilleifolia: variation in floral characteristics and outcrossing rate.
Evolution 36: 352-360.

Stebbins, G.L. 1950. Variation and Evolution in Plants.
Columbia University Press, New York.

Thomson, J.D. and S.C.H. Barrett. 1981. Selection for outcrossing, sexual selection, and the evolution of dioecy in plants.
Amer. Nat. 118: 441-449.

Uyenoyama, M.K. 1984. On the evolution of parthenogenesis: a genetic representation of the "cost of meiosis".
Evolution 38: 408-416.

Van Damme, J.M.M. 1983. Gynodieocy in Plantago lanceolata L. II. Inheritance of three male sterility types. Heredity 50: 253-274.

Van Damme. J.M.M. and W. Van Delden. 1982. Gynodioecy in Plantago lanceolata L. I. Polymorphism for plasmon type. Heredity 49: 303-318.

Westergaard, M. 1958. The mechanism of sex determination in dioecious flowering plants. Adv. Genet. 9: 217-281.

Williams, G.C. 1975. Sex and Evolution. Princeton University Press, Princeton.

Willson, M.F. 1979. Sexual selection in plants. Amer. Nat. 113: 777-790.

Willson, M.F. 1982. Sexual selection and dicliny in angiosperms. Amer. Nat. 119: 579-583.

EVOLUTION OF OUTBREEDING SYSTEMS

M.D. ROSS

ABSTRACT

This paper gives a personal view of the evolution of outbreeding systems, emphasizing the many features that hermaphrodite and monoecious populations, especially the more outbred ones, have in common with sex-polymorphic populations. Such features include frequency-dependency for fitness, functional sex, and combined gamete selfing rate. This last selfing rate differs from that for the ovules, since the pollen selfing rate differs from the ovule rate and is frequency dependent, because of variation in ovule and pollen fertilities. Five topics are emphasized. These are: (1) the concept of successful gametes, defined as gametes (of both sexes) that take part in fertilization; (2) sexual asymmetry, defined as non-constant pollen to ovule ratios among individuals (or genotypes or phenotypes); (3) allocation of resources to male or female reproduction, where such allocation may differ among individuals; (4) individual selection; and (5) genetic control. The concept of successful gametes is used to define fitness (the total number of successful gametes per individual), functional sex (the number of successful ovules as a proportion of all successful gametes, or fitness value), and combined gamete selfing rate (the number of ovules and pollen grains of an individual that take part in selfing, as a proportion of all successful gametes of that individual). Sexual asymmetry is held to be of fundamental importance not only for sex-polymorphic, but also for the more outbred hermaphrodite populations, and evidence for asymmetry is presented for hermaphrodite and monoecious populations. Asymmetry is responsible for maintenance of polymorphisms, and also for frequency-dependent selection, differential ovule and pollen outcrossing rates, and variation in extents of male and female functioning, so that these last three characters are common to asymmetric hermaphrodite and to sex-polymorphic populations. A simple model of male/female resource allocation ensures that any variation in such allocation results in asymmetry with frequency-dependent selection. Numerical examples show that intermediate or incompletely dominant gene action for resource allocation may result in overdominance or in equal genotypic fitnesses in equilibrium populations, for hermaphrodite, gynodioecious, or subdioecious populations. There is a continuum among these population types. It is held that individual selection is sufficient to bring about the evolution of sex polymorphisms, and such evolution is not always accompanied by the evolution of outcrossing. The importance of the mode of genetic control of a sex phenotype is emphasized, since differences in such control may entirely determine whether a polymorphism can be maintained or not.

INTRODUCTION

The evolution of outbreeding systems in seed plants has been intensively studied in the last decade or more. The purpose of the present paper is to give a personal view of the insights which may be obtained from these studies, rather than to provide a comprehensive review of the literature.

There are five points which seem to be of fundamental importance, namely: (1) *the concept of successful gametes* (2) *sexual asymmetry* (3) *male/female resource allocation* (4) *individual selection* and (5) *genetic control.* These points will be considered in this order, and an attempt is made to consider them in an evolutionary context, and to indicate their relevance not only for sex-polymorphic but also for hermaphrodite populations.

(1) THE CONCEPT OF SUCCESSFUL GAMETES

In this section successful gametes are first defined, and the application of this idea for obtaining evolutionarily important measures such as selfing rate and fitness is outlined. These measures are then obtained in detail, and applied in the later sections of this paper.

A gamete is regarded as successful if it takes part in fertilization, and this concept of successful gametes has proved extremely fruitful for measuring characters of evolutionary importance. In particular, fitness has been defined as the number of successful gametes per individual (*e.g.* Gregorius and Ross 1981). The degree to which a plant functions as female (Ross and Weir 1975) is then the number of its successful ovules as a proportion of all its successful gametes (or fitness value) and is called "functional sex" (Ross 1982; Ross and Gregorius 1983). The last two papers cited also present a new method for measuring selfing, which again uses the concept of successful gametes. The new selfing rate is called the "combined selfing rate" and is defined as the number of successful gametes (both ovules and pollen) which take part in selfing, as a proportion of all successful gametes (*i.e.* as a proportion of the fitness value). This new definition is needed because in general in populations which vary in fertility (see next section) the pollen selfing rate is different from the ovule rate. For example, the pollen selfing rate is lower in plants with more pollen than in those with less (Ross 1977a, 1977b, 1982; Ross and Gregorius 1983).

TABLE 1 A monogenic model of fertility variation

Character	Genotype A_1A_1	A_1A_2	A_2A_2
Female fertility	ϕ_{11}	ϕ_{12}	ϕ_{22}
Male fertility	μ_{11}	μ_{12}	μ_{22}
Ovule selfing rate	σ_{11}	σ_{12}	σ_{22}
Frequency	P_{11}	P_{12}	P_{22}

We now obtain for a simple genetic model the three quantities just considered. We assume a population with non-overlapping generations with a gene locus A and alleles A_1 and A_2. The locus controls male and female fertility, and also the ovule selfing rate. Genotype A_1A_1

produces ϕ_{11} ovules and μ_{11} pollen grains, and has ovule selfing rate σ_{11} and frequency P_{11}, and so on (Table 1). All non-selfed ovules are fertilized at random, and the amount of pollen used in selfing is taken to be negligible.

(a) Fitness values

We now obtain the number of successful gametes per individual (fitness value) of any genotype A_iA_j. The number of successful ovules from selfing equals $\sigma_{ij}\phi_{ij}$, and the number from crossing equals $(1-\sigma_{ij})\phi_{ij}$. The number of successful pollen grains from selfing also equals $\sigma_{ij}\phi_{ij}$. The number of successful pollen grains from crossing equals the number of crossed ovules in the population, $\overline{\phi(1-\sigma)}$, multiplied by the genotype A_iA_j's pollen fertility (μ_{ij}) as a proportion of the total population pollen production, $\bar{\mu}$. In these expressions we have

$$\overline{\phi(1-\sigma)} = (1-\sigma_{11})\phi_{11}P_{11} + (1-\sigma_{12})\phi_{12}P_{12} + (1-\sigma_{22})\phi_{22}P_{22}$$

and

$$\bar{\mu} = \mu_{11}P_{11} + \mu_{12}P_{12} + \mu_{22}P_{22}$$

If the P's are taken to represent proportions, instead of actual numbers, then $\overline{\phi(1-\sigma)}$ and $\bar{\mu}$ represent mean number of crossed ovules and of pollen grains per plant in the population, respectively.

We now add these four quantities, and obtain the total number of sucessful gametes per individual, or fitness value

$$w_{ij} = (1+\sigma_{ij})\phi_{ij} + \overline{\phi(1-\sigma)}\,\mu_{ij}/\bar{\mu}$$

These fitness values are frequency-dependent, since they contain the frequency-dependent quantities $\overline{\phi(1-\sigma)}$ and $\bar{\mu}$. This reflects the assumption that the number of pollen grains successful in crossing depends upon the number of ovules available for crossing. The number of crossed ovules depends in turn on the ovule crossing rates and ovule fertilities of the various genotypes, and therefore on the genotype frequencies. Similarly, because of pollen competition, the number of pollen grains successful in crossing depends also on the various genotypic pollen fertilities, and therefore on the genotype frequencies. Of course, this frequency dependency occurs only if there is genotypic variation in crossing rates or fertilities. See Gregorius and Ross (1981) for a more detailed discussion of this fitness concept. This concept has been extended by Gregorius (1984b; 1984c), and a study where selection resulted from different genotypic selfing rates alone yielded precise conditions for maintaining polymorphisms (Gregorius 1984d). Such selection allowed overdominance or underdominance, for example. A polymorphism occurred for some cases of overdominance or underdominance, but global convergence to a unique polymorphic equilibrium required overdominance.

This fitness model has led to powerful methods for studying the maintenance of polymorphisms and for obtaining equilibrium frequencies. For example, one can easily obtain equilibrium frequencies for a dominant model even if male fertility, female fertility, viability and ovule selfing rates all differ genotypically (see section 2).

(b) Functional sex

The degree to which a plant functions as male or female depends not only on its ovule or pollen fertility, but also on the frequencies of the other types in the population. For example, if a particular genotype occurs in a population consisting of that genotype alone, then it of necessity reproduces half as female and half as male, *i.e.* it has a functional sex of 0.5. If that same genotype occurs in a population consisting largely of equally ovule-fertile, but much more pollen-fertile types, then it will reproduce largely as female, *i.e.* it will have a functional sex of more than 0.5. This is because of pollen competition. The frequency dependency is reflected in the expression for functional sex of any genotype $A_i A_j$

$$F_{ij} = \phi_{ij}/w_{ij}$$

It has already been shown that the frequency dependency of the fitness values (w_{ij}) may result, *inter alia,* from different ovule selfing rates and different fertilities among genotypes (Ross 1984b). Therefore, since the expression for functional sex contains the fitness value w it is clear that these factors cause frequency dependency for functional sex also. Nevertheless, it may be interesting to look at the effects of different ovule selfing rates on functional sex, so that the close connection between these measures becomes apparent. We set all ϕ's and μ's equal, so that $\phi_{ij}=\bar{\phi}=\phi$ and $\mu_{ij}=\bar{\bar{\mu}}=\mu$. This yields

$$F_{ij} = 1/(2 + \sigma_{ij} - \sigma_{11}P_{11} - \sigma_{12}P_{12} - \sigma_{22}P_{22})$$

so that the functional sex depends upon the frequencies as long as not all ovule selfing rates are the same. When they are all equal, including the cases of all equal to zero or one, functional sex is no longer frequency dependent, and equals one half. It seems very unlikely that real partially-selfing populations should show equal genotypic fertilities and selfing rates. For a dominant gene with $\sigma_{11} = \sigma_{12}$ the connection between the frequency dependency and the different selfing rates is even more apparent. The functional sex for genotype A_1A_1, for example, is then

$$1/[2 + P_{22}(\sigma_{11} - \sigma_{22})]$$

which equals 1/2 for $\sigma_{11} = \sigma_{22}$. If $\sigma_{11} = 0$ and $\sigma_{22} = 1$, the functional sex of genotype A_1A_1 = $1/(2 - P_{22})$, and is therefore an increasing function of P_{22}. Thus genotype A_1A_1 functions more as female as the frequency of A_2A_2 increases. This is so because the pollen of A_1A_1 has fewer and fewer ovules available (for crossing, since none of its pollen takes part in selfing and none of the ovules of A_2A_2 are crossed). F_{11} is an increasing function of P_{22} for all $\sigma_{11} < \sigma_{22}$.

(c) Combined selfing rate

The combined selfing rate is the number of gametes (of both sexes) which take part in selfing as a proportion of all successful gametes, and thus equals

$$S_{ij} = 2\sigma_{ij}\phi_{ij}/m_{ij}$$

It thus equals twice the ovule selfing rate multiplied by the functional sex, which shows its close relationship to functional sex. The equation also demonstrates the close relationships between fitness, sex function, male and female fertility, selfing rates and genotype frequencies.

Studies of real populations show that the combined selfing rate can differ considerably from the ovule rate (see Ross 1984a and below), and the combined selfing rate is to be preferred for any population which varies in fertility or ovule selfing rate.

We now look at the effects of some of the components of this selfing rate. If we set all σ's equal, so that $\sigma_{ij}=\sigma$, then

$$\overline{\phi}_{(1-\sigma)} = (1-\sigma)(\phi_{11}P_{11} + \phi_{12}P_{12} + \phi_{22}P_{22}) = (1-\sigma)\overline{\phi}$$

so that

$$S_{11} = 2\sigma\,\phi_{11}/[(1+\sigma)\phi_{11} + (1-\sigma)\overline{\phi}\mu_{11}/\overline{\mu}]$$

Thus the combined selfing rate remains different for the different genotypes, and depends on the male and female fertilities and on frequencies. The combined rate is different from the ovule rate, since

$$S_{11}-\sigma = [2\sigma\phi_{11} - \sigma[(1+\sigma)\phi_{11} + (1-\sigma)\overline{\phi}\mu_{11}/\overline{\mu}]]/[(1+\sigma)\phi_{11} + (1-\sigma)\overline{\phi}\mu_{11}/\overline{\mu}]$$

which easily reduces to

$$\sigma(1-\sigma)(\phi_{11} - \overline{\phi}\mu_{11}/\overline{\mu})/[(1+\sigma)\phi_{11} + (1-\sigma)\overline{\phi}\mu_{11}/\overline{\mu}]$$

Since the denominator > 0, $S_{11} - \sigma \neq 0$ provided $0 < \sigma < 1$ and $\phi_{11} \neq \overline{\phi}\mu_{11}/\overline{\mu}$, *i.e.* $\phi_{11}/\mu_{11} \neq \overline{\phi}/\overline{\mu}$. Thus for partially selfing populations the combined selfing rate of type A_1A_1 differs from the ovule selfing rate as long as the ratio of this genotype's pollen : ovule fertility does not equal that of the population mean. This is the case if the population is not monomorphic for A_1A_1 and if there is asymmetric variation for ovule or pollen fertility. Asymmetry is the case where there is population variation in pollen : ovule fertility ratios (Ziehe and Gregorius 1981; section 2 of the present paper).

If we retain genotypically variable σ's, but now set all $\phi_{ij} = \phi$, we obtain

$$S_{11} = 2\sigma_{11}/[1+\sigma_{11} + (1-\sigma_{11}P_{11} - \sigma_{12}P_{12} - \sigma_{22}P_{22})\mu_{11}/\overline{\mu}]$$

so that the combined selfing rate of A_1A_1 depends upon the ovule selfing rates, and on the frequencies and pollen fertilities of all genotypes in the population. Similarly, it is easy to show that if all μ's only are set equal then S_{11} depends on ovule selfing rates, genotype frequencies and ovule fertilities. For all ϕ's and μ's set equal

$$S_{11} = 2\sigma_{11}/[2 + P_{12}(\sigma_{11}-\sigma_{12}) + P_{22}(\sigma_{11}-\sigma_{22})]$$

So that the combined selfing rate depends on the various genotypic ovule selfing rates and frequencies. For dominant A_1 and therefore $\sigma_{11} = \sigma_{12}$

$$S_{11} = S_{12} = 2\sigma_{11}/[2 + P_{22}(\sigma_{11}-\sigma_{22})]$$

so that for $\sigma_{11} > \sigma_{22}$, for example, increased P_{22} results in decreased S_{11}. The combined selfing rate of the genotypes with the greater ovule selfing rate decreases as the frequency of the other genotype increases. This is because as the other genotype increases in frequency, more ovules become available for crossing in the population.

Similarly, it is clear for all σ's and ϕ's or all σ's and μ's equal, that the combined selfing rate depends on frequencies and on μ's or ϕ's, respectively.

(2) SEXUAL ASYMMETRY

The situation where pollen:ovule ratios are not constant among phenotypes or genotypes in a population is called sexual asymmetry, and the converse case of constant ratios sexual symmetry (Ziehe & Gregorius 1981). This concept is perhaps the most fundamental and important underlying the whole field of the evolution of outbreeding systems in seed plants. It is important not only for understanding the origin of sex polymorphisms such as gynodioecy (presence of female and hermaphrodite plants) or dioecy, but is perhaps even more important for a proper understanding of monomorphic populations, such as hermaphrodite or monoecious ones. This is because of the tacit assumption often made that, for example, hermaphrodite populations are uniform in their sex expression. In a sense this assumption is correct, since by definition all of the plants are hermaphrodite. But the further assumption that all plants in a hermaphrodite population are equally male and female fertile finds no general confirmation in the literature, and is intuitively rejected (at least for non-inbreeders) as soon as it is made explicit. Theoretically, fertility variants could all show proportionate increases (or decreases) in both pollen and ovule fertility relative to some standard type, so that the population remains sexually symmetric with constant pollen:ovule ratios despite the occurrence of fertility variation. This, however, seems unlikely both intuitively (Ross 1977b) and on grounds of resource allocation theory (see section 2 and Ross and Gregorius 1983), and is apparently not found in real populations (*e.g.* Primack 1979; Ross 1984b). A hermaphrodite has two ways in which it can show variation in fertility, namely as male and as female, and it seems extremely unlikely that a variant should differ from some standard type to exactly the same extent in each sex function (Ross 1977b). Fertility variation, like other forms of variation, is only to be expected, and in hermaphrodite populations such variation will in practice always result in sexual asymmetry. Exactly the same considerations apply to monoecious populations, and for the rest of this article both types of population will be subsumed under the term "hermaphrodite".

We now consider some effects of sexual asymmetry (hereafter called simply "asymmetry"). Perhaps the most important effect of asymmetry is that it may allow the maintenance of a polymorphism for a monogenic dominant model (Gregorius 1984a), where such maintenance is not possible under the classical model of viability selection. Such maintenance is also not possible under a model of symmetrical fertility selection, which is equivalent to the classical viability model. Gregorius has used a new method to show that "antisymmetry", defined as the situation where increased male is associated with reduced female fertility, and vice versa, is a necessary condition for a polymorphism to occur under a dominance model. Clearly, antisymmetry is a special case of asymmetry. It is also possible to obtain this result by using the fitnesses. We use the model in Table 1, with A_1 dominant to A_2, so that $\phi_{11} = \phi_{12}$, $\mu_{11} = \mu_{12}$, and $\sigma_{11} = \sigma_{12}$. Allele A_1 will increase in frequency in populations consisting almost entirely of A_2A_2 if $w_{11} > w_{22}$ as P_{22} approaches 1. At this frequency $\overline{\phi(1-\sigma)} = (1-\sigma_{22})\phi_{22}$ and $\bar{\mu} = \mu_{22}$ so that $w_{22} = 2\phi_{22}$. This yields the required inequality

$$(1+\sigma_{11})\phi_{11} + (1-\sigma_{22})\phi_{22}\mu_{11}/\mu_{22} > 2\phi_{22}$$

Similarly, A_2 increases in frequency as P_{11} approaches 1 if

$$(1+\sigma_{22})\phi_{22} + (1-\sigma_{11})\phi_{11}\mu_{22}/\mu_{11} > 2\phi_{11}$$

Multiplying the first inequality by μ_{22} and the second by μ_{11}, adding the inequalities and rearranging yields the following condition for the maintenance of a polymorphism

$$(\phi_{11} - \phi_{22})(\mu_{22} - \mu_{11}) > 0 \qquad (1)$$

This requires $\phi_{11} > \phi_{22}$ and $\mu_{11} < \mu_{22}$, or $\phi_{11} < \phi_{22}$ and $\mu_{11} > \mu_{22}$, and therefore requires $\phi_{11} \neq \phi_{22}$, $\mu_{11} \neq \mu_{22}$. Thus one homozygote must have more ovules and less pollen than the other (antisymmetry).

If we apply a resource-allocation model (section 3) to this result, we find that this condition reduces to $\phi_{11} \neq \phi_{22}$. This is because the model assumes equal reproductive resources for each genotype, so that the type with fewer ovules of necessity has more pollen, and vice versa. Writing r^f for the amount of resources required for one ovule, and r^m for the amount required for one pollen grain, we see that genotype A_1A_1 uses $\phi_{11}r^f$ of its resources for ovules, leaving $1-\phi_{11}r^f$ for pollen grains, of which $(1-\phi_{11})r^f/r^m$ are produced. Substituting such μ values into inequality (1) yields

$$(\phi_{11} - \phi_{22})^2 r^f/r^m > 0$$

which requires $\phi_{11} \neq \phi_{22}$. Thus for simple and biologically meaningful assumptions any fertility variation could result in a polymorphism.

The above result illustrates something of the power of the fitness method. Since the fitness values also include viabilities from immediately after zygote formation in one generation to the corresponding time in the next (Gregorius and Ross 1981), we see that we can obtain analytical results which allow for variation in viability, male and female fertility and selfing rates. Further developments in this area have led, for example, to analytical results for situations without dominance (Gregorius 1982; 1984b;1984c). For a dominant gene, we may obtain polymorphic equilibrium frequencies by setting $w_{11} = w_{22}$, yielding

$$\hat{P}_{22} = [\phi_{11}\mu_{22}(1-\sigma_{11}) + \phi_{22}\mu_{11}(1+\sigma_{22}) - 2\phi_{11}\mu_{11}]/2(\phi_{11}-\phi_{22})(\mu_{22}-\mu_{11})$$

For $\sigma_{11} = \sigma_{22} = \sigma$, this reduces to

$$\hat{P}_{22} = [(1-\sigma)\phi_{11}\mu_{22} + (1+\sigma)\phi_{22}\mu_{11} - 2\phi_{11}\mu_{11}]/2(\phi_{11}-\phi_{22})(\mu_{22}-\mu_{11})$$

and for $\sigma = 0$ to

$$\hat{P}_{22} = [(\phi_{22}-\phi_{11})\mu_{11} + (\mu_{22}-\mu_{11})\phi_{11}]/2(\phi_{11}-\phi_{22})(\mu_{22}-\mu_{11})$$

Numerical results show that this equilibrium is globally attractive and stable.

As an application of this result to a real population we may use the example studied by Ross and Gregorius (1983). In a race of *Leavenworthia crassa* polymorphic for flower colour, Lloyd (1965) found that the dominant type with yellow-centred flowers had 3099 ovules and 21,412,000 pollen grains. The corresponding figures for the recessive with yellow flowers were 3503 and 17,695,200. Assuming two alleles and equal σ's, these figures lead to an equilibrium frequency for the recessive of 0.388 for $\sigma = 0.2$, and to 0.724 for $\sigma = 0.25$. Notice the extreme sensitivity of the equilibrium frequency to the selfing rate. For $\sigma \leq 0.142$ the population becomes monomorphic for the dominant type, and for $\sigma \geq 0.291$ it becomes monomorphic for the recessive.

TABLE 2 Asymmetry, functional sex and fitness in Scots Pine (*Pinus sylvestris*) (Ross 1984a)

Clone	ϕ	μ	μ/ϕ	Functional sex	Fitness
CH 10	500.33	48,335	97	0.782	640
RD 8	223.50	13,487	60	0.825	271
CH 11	171.67	28,586	167	0.170	242
CH 18	197.00	60,261	306	0.601	328
CH 9	160.75	31,005	193	0.687	234
CH 13	85.00	58,650	690	0.422	202
CH 1	83.50	17,916	215	0.668	125
RD 1	80.25	14,075	175	0.702	114
CH 20	33.50	3,946	118	0.759	44
CH 4	26.25	53,434	2036	0.206	127
CH 19	13.75	99,829	7260	0.069	199
CH 2	13.50	68,852	5100	0.095	142
CH 7	11.67	11,172	958	0.349	33

Another application in this area involves a seed-plantation of clones of Scots Pine, *Pinus sylvestris* (for further details see Ross 1984b). The number of male and female cones per clone (genotype) was estimated, and the ovule selfing rate had previously been estimated as 0.101 (Müller-Starck, pers. comm.). Table 2 gives results for some of the clones and shows that they varied greatly in numbers of male cones (μ) and female cones (ϕ). The ratio μ/ϕ, which is a measure of the asymmetry, varied immensely (from 97 to 7260), and it seems very probable that asymmetry is of significant importance in this population. These figures lead to estimates for functional sex varying from a low of 0.068 (nearly male) to a high of 0.825 (predominantly female), with the other clones showing a range of intermediate values (Ross 1984a). No clone in this nominally monoecious species showed 50% female and 50% male function. Extreme fitness values differed from each other by a factor of about 19. Similar fitness values could be reached in different ways. For example, the most nearly male clone with functional sex 0.069 had a fitness of 199, whereas a clone which came near to functioning half as female and half as male (functional sex 0.422) had a fitness of 202 (Ross 1984a). Although an ovule selfing rate of 0.101 was assumed to apply to all genotypes, combined gamete selfing rates varied among genotypes and differed from the ovule rates. Thus a clone with weak ovule but fairly strong pollen production had a combined selfing rate of 0.167, whereas another clone with the reversed type of gamete production had a rate of 0.019 (Ross 1984a).

So far we have considered asymmetry in hermaphrodite populations only. One aim of this paper, however, is to suggest that hermaphrodite populations are usually asymmetrical, and differ from sex-polymorphic populations only in degree. The pine population, with its predominantly female genotypes and near-male ones, in addition to its less asymmetrical genotypes, is fundamentally similar to a trioecious population, consisting of males, females and hermaphrodites, or to a gynodioecious population, having females and hermaphrodites. All of the features so far considered, such as frequency-dependent fitness, different pollen and ovule selfing rates, and functional sex $\neq$ 0.5 are found in sex-polymorphic populations also. For example, in gynodioecious populations the pollen outcrossing rate of the (partially selfing) hermaphrodites is greater than their ovule rate (Ross 1977b) for two reasons. First, the pollen of the hermaphrodites pollinates all the ovules in the population, and not only those of the hermaphrodites, and second the females often have more ovules or seeds than the hermaphrodites. For the same reasons, the hermaphrodites function more as male than as female parents (Lloyd 1974). We may see the close relationship between gynodioecious and asymmetrical hermaphrodite populations by considering fitnesses and the closely related parameters of functional sex and combined selfing rates. We adopt the monogenic model given in Table 1, and set genotype A_2A_2 as the female genotype, so that $\mu_{22} = \sigma_{22} = 0$. The other two genotypes are hermaphrodites, yielding a model of monogenic recessive gynodioecy. Using the methods already given, we find that the fitnesses of the two hermaphrodite genotypes can be written in exactly the same way as for the hermaphrodite populations already considered. However, $\overline{\phi(1-\sigma)}$ now equals

$$(1-\sigma_{11})\phi_{11}P_{11} + (1-\sigma_{12})\phi_{12}P_{12} + \phi_{22}P_{22}$$

and $\bar{\mu}$ equals $\mu_{11}P_{11} + \mu_{12}P_{12}$. The fitnesses of the hermaphrodites remain frequency-dependent, but that of the female is constant and equals ϕ_{22}. If the model were to allow for reduced seed set on females with increased female frequency, because of limitations in pollen supply, for example, then female fitnesses could also be frequency-dependent. Notice that dominance in the model refers only to hermaphroditism or femaleness, and does not require the hermaphrodites to be equally fertile or to have the same ovule selfing rates. If we make these additional assumptions, so that A_1 becomes dominant for all of the model parameters, then we have $\phi_{11} = \phi_{12}$, $\sigma_{11} = \sigma_{12} = \sigma$,

$$\bar{\phi}(1-\sigma) = (1-\sigma)\phi_{11}(1-P_{22}) + \phi_{22}P_{22}$$

and $\bar{\mu} = \mu_{11}(1-P_{22})$. Females are maintained in the population if $w_{22} > w_{11}$ as P_{11} approaches 1. At this frequency, $\bar{\phi}(1-\sigma) = (1-\sigma)\phi_{11}$ and $\bar{\mu} = \mu_{11}$, so that $w_{11} = 2\phi_{11}$. Thus females are maintained if $\phi_{22} > 2\phi_{11}$, *i.e.* if females have more than twice the seed fertility of the hermaphrodites, which is the old result of Lewis (1941). We assume equal genotypic resources, and let the total amount of resources used for ovule production equal $R_{11} = \phi_{11}r^f$ for genotype A_1A_1 and so on for the other genotypes (see Table 3 and Ross and Gregorius 1983). Therefore, since all of the resources of A_2A_2 are used for ovules, R_{22} equals one. Thus $\phi_{22} > 2\phi_{11}$ requires $R_{22} > 2R_{11}$, which requires $R_{11} < 0.5$, so that females can only be maintained if the hermaphrodites put less than half their resources into ovule production. This result can also be obtained by applying the results of Gregorius (1984a). For A_1 dominant in all respects

$$w_{11} = w_{12} = 2\phi_{11} + \phi_{22}P_{22}/(1-P_{22})$$

so that the fitness of the hermaphrodites always increases with the frequency of the females, *i.e.* decreases with hermaphrodite frequency and so is negatively frequency dependent. At equilibrium $w_{11} = w_{22}$, which yields $\hat{P}_{22} = (\phi_{22}-2\phi_{11})/2(\phi_{22}-\phi_{11})$, or, expressed in terms of resource allocation, $\hat{P}_{22} = (1-2R_{11})/2(1-R_{11})$. Thus, for example, if the hermaphrodites put one third of their reproductive resources into ovules, then we obtain an equilibrium frequency for females of 0.25. For functional sex of A_1A_1 (F_{11}) we obtain $F_{11} = \phi_{11}(1-P_{22})/[2\phi_{11} + P_{22}(\phi_{22}-2\phi_{11})]$, which for the model of resource allocation equals

$$R_{11}(1-P_{22})/[2R_{11} + P_{22}(1-2R_{11})]$$

Thus as the number of females P_{22} increases, the functional sex of the hermaphrodites decreases, so that they behave more as male parents. Similarly, the combined selfing rate of the hermaphrodites, which is simply 2σ times their functional sex, also decreases as the number of females increases. In summary, we see that asymmetric hermaphrodite populations and gynodioecious populations (which are by definition asymmetric) show many common characteristics, such as frequency-dependent selection, for example. These features are also found in other sex-polymorphic systems, such as trioecy or subdioecy.

(3) MALE/FEMALE RESOURCE ALLOCATION

Both asymmetry in hermaphrodites and the evolution of sex-polymorphisms involve the distribution of resources between male and female functions, so that for this reason alone resource-allocation theory is of fundamental importance in the present context. This section, however, is concerned with the application of such theory to answering two questions only. First, what is the influence of resource allocation on fitnesses, and second, can resource-allocation theory be used for obtaining realistic ovule and pollen fertility values for use in our models? With respect to the second question, if we take, for example, very high male and female fertility values for a particular genotype in a model with dominance, and assume lower fertilities for the other genotypes, then it is not surprising if the the first genotype displaces the others from the population. Such a result does not bring any new knowledge, and the situation is probably also not meaningful biologically. Similarly, if we set a high enough ovule fertility for females in an initially hermaphrodite population, it can come as no surprise to find that the females can establish themselves. Therefore, one important assumption of the present resource-allocation model is that the various genotypes have equal reproductive resources, and differ only in how these resources are distributed between male and female functions. In this way we hope to obtain realistic fertility values, and can also study the effects of the allocation of resources without confounding such effects with those resulting from variation in the total amounts of resources available. It would not be surprising if there were variation in total reproductive resources available to different genotypes in natural populations, and such a situation also deserves to be studied.

The essential features of the resource-allocation model (Table 3) have already been presented in the previous section. The total reproductive resources of each genotype are set

TABLE 3 A monogenic model of resource allocation (Ross and Gregorius 1983)

Character	Genotype A_1A_1	A_1A_2	A_2A_2
Ovule fertility	ϕ_{11}	ϕ_{12}	ϕ_{22}
Pollen fertility	μ_{11}	μ_{12}	μ_{22}
Resources used for ovules	R_{11}	R_{12}	R_{22}
Resources used for pollen	$1-R_{11}$	$1-R_{12}$	$1-R_{22}$
Ovule selfing rate	σ_{11}	σ_{12}	σ_{22}
Frequency	P_{11}	P_{12}	P_{22}

equal to one, and of this amount proportion r^f is used for one ovule and r^m for one pollen grain. Thus genotype A_iA_j uses $\phi_{ij}r^f = R_{ij}$ of its resources for ovules, leaving $1-\phi_{ij}r^f = 1-R_{ij}$ available for pollen, with the result that $(1-R_{ij})/r^m$ pollen grains are produced. The remaining features of the model are as given in Section 1 and Table 1.

Some answers to the first question, concerning the effects of resource allocation on fitness, were obtained numerically, and are given in Tables 4 and 5. Analytical studies of a similar model, but with all σ's equal, are given by Ziehe (1985). Table 4 gives the results for two equilibrium hermaphrodite populations, which, because of resource-allocation model, are necessarily asymmetric. Both populations have R values of 0.4 (for A_1A_1) 0.5 (A_1A_2) and 0.6 (A_2A_2), so that there is intermediate gene action for resources and therefore for fertilities. However, the respective ovule selfing rates are 0.1, 0.1, 0.1 for the first, and 0.1, 0.2, 0.1 for

TABLE 4 Equilibrium results for asymmetric hermaphrodite populations

Parameter	Genotype		
	A_1A_1	A_1A_2	A_2A_2
Population 1			
Resources for ovules	0.4	0.5	0.6
Ovule selfing rate	0.1	0.1	0.1
Frequency	0.070	0.360	0.570
Fitness	1.100	1.100	1.100
Functional sex	0.364	0.455	0.545
Combined selfing rate	0.073	0.091	0.109
Population 2			
Resources for ovules	0.4	0.5	0.6
Ovule selfing rate	0.1	0.2	0.1
Frequency	0.135	0.438	0.428
Fitness	1.019	1.083	1.046
Functional sex	0.322	0.394	0.502
Combined selfing rate	0.064	0.158	0.100

the second population. Despite the intermediate gene action, the first population shows equal genotypic fitnesses, and the second overdominance. Non-equilibrium fitnesses were intermediate in the first, and overdominant in the second population. As expected, functional sex increased with the resources devoted to ovules, as did the combined selfing rate (2σ times functional sex). Notice the considerable effects of a slight change in ovule selfing rate on the equilibrium frequency, fitness (including mean population fitness) and functional sex.

Table 5 gives results for gynodioecious populations, where the gene action is again intermediate for resource allocation (R values of 0.3, 0.65 and 1), but where fitnesses again show overdominance. Notice that gynodioecy is recessive with respect to hermaphroditism, and that the ovule selfing rate is also recessive. Despite equal ovule selfing rates for the two hermaphrodite genotypes, the heterozygous hermaphrodite has the greater combined selfing rate. One could regard the population as showing recessiveness for gynodioecy and ovule selfing rate, intermediate gene action for resource allocation and functional sex, and overdominance for fitness and combined selfing rate. Another value for resource allocation (0.5, 0.75, 1) gave similar results.

TABLE 5 Equilibrium results for gynodioecy

Parameter	Genotype		
	A_1A_1	A_1A_2	A_2A_2
Resources for ovules	0.3	0.65	1.0
Ovule selfing rate	0.2	0.2	0.0
Frequency	0.406	0.484	0.109
Fitness	1.067	1.134	1.000
Functional sex	0.281	0.573	1.000
Combined selfing rate	0.112	0.229	0.000

Table 6 gives results for subdioecious populations, where subdioecy is defined as the occurrence of imperfectly differentiated individuals in addition to perfectly differentiated (strictly unisexual) ones. Subdioecy occurs frequently in seed plants, even in "dioecious" species which are regarded as well differentiated, such as many poplars (Lester 1963). Population 1 in Table 6 has resource-allocation values of 0.1, 0.2, and 1, so that the A_2A_2 or XX females are well differentiated, whereas the A_1A_2 or XY males and the A_1A_1 or YY males are imperfectly differentiated, and produce some ovules. Such a situation occurs relatively frequently in subdioecious plants thought to have gynodioecious ancestors (Ross 1982). In population 1 the Y chromosome acts as a single dominant gene with respect to male vs female sex expression, but is incompletely dominant for resource allocation and therefore for the quantities of ovules and pollen produced. Such a situation leads to overdominance in equilibrium populations (Table 6, and Ross 1982 for real populations). Notice that the real selfing rate of the males is considerably lower than their ovule rate. In population 2 the Y chromosome is dominant for both male sex expression and resource allocation, and all three fitnesses are therefore equal in equilibrium populations (Gregorius 1981).

TABLE 6 Equilibrium results for subdioecy

Parameter	Genotype		
	A_1A_1	A_1A_2	A_2A_2
Population 1			
Resources for ovules	0.1	0.2	1.0
Ovule selfing rate	0.3	0.3	0.0
Frequency	0.068	0.545	0.387
Fitness	0.976	1.012	1.000
Functional sex	0.102	0.198	1.000
Combined selfing rate	0.061	0.119	0.000
Population 2			
Resources for ovules	0.1	0.1	1.0
Ovule selfing rate	0.3	0.3	0.0
Frequency	0.031	0.524	0.444
Fitness	1.000	1.000	1.000
Functional sex	0.100	0.100	1.000
Combined selfing rate	0.060	0.060	0.000

This section may perhaps be summarised by emphasizing that there is a continuum between asymmetrical hermaphrodite and sex-polymorphic populations. Hermaphrodite populations may have genotypes that are strongly differentiated towards male or female functions, and subdioecious or even "dioecious" populations may be imperfectly differentiated.

Such populations have many characteristics in common, such as frequency-dependent fitness, functional sex and combined selfing rate. Combined selfing rates differ among genotypes, even when ovule selfing rates are the same.

We have seen that resource-allocation theory may result in overdominance or equal genotypic fitnesses in equilibrium hermaphrodite or sex-polymorphic populations, even when the distribution of reproductive resources is controlled by intermediate or incompletely dominant gene action (see also Ziehe 1985). The second question regarding resource-allocation theory was whether such theory could be used to obtain realistic fertility values for use in models of evolution of sex polymorphisms? For example, when hermaphroditism is dominant, and both hermaphrodite genotypes have equal fertilities and ovule selfing rates, females require a more than doubled ovule fertility to establish themselves. Is such a high fertility value likely? For such an initial hermaphrodite population, the optimum resource allocation equals $1/2(1+\sigma)$, and was defined as the allocation which resulted in an equilibrium monomorphic for the type having it (Ross and Gregorius 1983). In such a population the females could not establish themselves, even for $\sigma=0$. However, we have seen that there is good evidence for regarding hermaphrodite populations as asymmetric and polymorphic, so we may ask if the females have a better chance in such populations? This question is more complicated, and may require a three-allele model. However, we may obtain some answer of interest by using the results of Gregorius, Ross and Gillet (1982). For recessive females and no selfing in the hermaphrodites, females (genotype A_2A_2) were maintained if $\phi_{12}/\phi_{11} + \mu_{12}/\mu_{11} > 2$. Replacing the ϕ's by R's and μ's by $(1-R)$'s, we obtain

$$(1-2R_{11})(R_{12}-R_{11}) > 0$$

This is fulfilled if $R_{11} \neq 0.5$ and $R_{11} \neq R_{12}$, and requires that if $R_{11} > 0.5$, then $R_{11} > R_{12}$, or if $R_{11} < 0.5$, then $R_{11} < R_{12}$. Thus $R_{11} = 0.3$ and $R_{12} = 0.65$ allows a polymorphism for σ's = 0 also. Numerical studies show that these conditions lead to equal fitnesses in equilibrium populations.

Another situation where R values were applied to gynodioecious populations is for nucleocytoplasmic control of gynodioecy, where for some modes of gene-cytoplasm interaction the present model of resource allocation always allowed the evolution of gynodioecy (Ross and Gregorius 1985). This case emphasizes the important effect of the mode of genetic control for the evolution of sex polymorphisms (Section 5). The previous case, where gynodioecy could evolve for σ's = 0 shows that gynodioecy need not always function as an outbreeding system (see next Section).

(4) INDIVIDUAL SELECTION

Earlier studies of the evolution of gynodioecy or subdioecy have considered the topic from the viewpoint of the evolution of outbreeding systems. It might be thought that such evolution could involve group selection, rather than individual selection. However, the occurrence of gynodioecy in self incompatible species in nature (Ross 1970; Horovitz and Galil 1972), and its apparently ready evolution in the models for σ's = 0, show that the evolution of gynodioecy need not involve the evolution of outbreeding. The situation for dioecy is less clear. The results with models show that individual selection alone is sufficient to allow the evolution of a sex polymorphism, which may or may not be associated with the evolution of outbreeding (Ross and Gregorius 1985). The mode of gene action is probably very important in this respect (see next Section). Many other, predominantly ecological, hypotheses for the evolution of dioecy have been presented (Bawa 1980).

(5) GENETIC CONTROL

It is entirely possible to invent one-locus two-allele models of sex polymorphisms where the polymorphism is always, sometimes or never established, depending entirely on the mode of genetic control (Table 7; Gregorius, Ross and Gillet 1982; 1983). Table 7 gives some examples for gynodioecy and trioecy. Similarly, the establishment of gynodioecy is much easier under models of gene-cytoplasm inheritance, compared to nuclear inheritance (Ross and Gregorius 1985). Gynodioecy seems unlikely to evolve further to give dioecy under gene-cytoplasm inheritance, whereas purely nuclear inheritance, at least if it is simple, apparently readily

TABLE 7 Mode of genetic control and maintenance of sex polymorphisms

Protected?	Genotype and phenotype A_1A_1	A_1A_2	A_2A_2
Always	Female	Hermaphrodite	Female
Sometimes	Hermaphrodite	Hermaphrodite	Female
Always	Female	Hermaphrodite	Male
Never	Female	Male	Hermaphrodite

evolves further towards dioecy (Ross 1978). It therefore seems that ESS models that do not refer to a particular genetic system, or models of inbreeding depression which do not specify the genetic basis of this phenomenon, may have only limited relevance to the evolution of outbreeding.

LITERATURE CITED

BAWA, K.S. (1980). Evolution of dioecy in flowering plants. Ann. Rev. Ecol. Syst. 11:15-39.

GREGORIUS, H.-R. (1981). Realized genotypic fitnesses at equilibrium in the deterministic selection theory of a diallelic locus. Göttingen Res. Notes Forest Genet. 4.

GREGORIUS, H.-R. (1982). Selection in plant populations of effectively infinite size. II. Protectedness of a biallelic polymorphism. J. Theor. Biol. 96: 689-705.

GREGORIUS, H.-R. (1984a). Selection with two alleles and complete dominance. Biol. J. Linn. Soc. 23: 157-165.

GREGORIUS, H.-R. (1984b). Fractional fitnesses in exclusively sexually reproducing populations. J. Theor. Biol. 111: 205-229.

GREGORIUS, H.-R. (1984c). Allele protectedness in frequency-dependent biallelic selection models with separated generations. J. Theor. Biol. 111: 425-446.

GREGORIUS, H.-R. (1984d). Convergence of genotypic frequencies for differential selfing and positive assortative mating at a biallelic locus. J. Math. Biol. 20: 159-169.

GREGORIUS, H.-R., and M.D. ROSS (1981). Selection in plant populations of effectively infinite size: I. Realized genotypic fitnesses. Math. Biosc. 54: 291-307.

GREGORIUS, H.-R., M.D. ROSS, and E. GILLET (1982). Selection in plant populations of effectively infinite size. III. The maintenance of females among hermaphrodites for a biallelic model. Heredity 48: 329-343.

GREGORIUS, H.-R., M.D. ROSS, and E. GILLET (1983). Selection in plant populations of effectively infinite size. V. Biallelic models of trioecy. Genetics 103: 520-544.

HOROVITZ, A., and J. GALIL (1972). Gynodioecism in east Mediterranean *Hirschfeldia incana*. Cruciferae. Bot. Gaz. 133: 127-131.

LESTER, D.T. (1963). Variation in sex expression in *Populus tremuloides* Michx. Silvae Genet. 12: 141-151.

LEWIS, D. (1941). Male sterility in natural populations of hermaphrodite plants. New Phytol. 40: 56-63.

LLOYD, D.G. (1965). Evolution of self-compatibility and racial differentiation in *Leavenworthia* (Cruciferae). Contrib. Gray Herb. Harvard Univ. 195: 3-134.

LLOYD, D.G. (1974). Theoretical sex ratios of dioecious and gynodioecious Angiosperms. Heredity 32: 11-34.

PRIMACK, R.B. (1979). Reproductive biology of *Discaria toumatou* (Rhamnaceae). N.Z. J. Bot. 17: 9-13.

ROSS. M.D. (1970). Breeding systems in *Plantago*. Heredity 25: 129-133.

ROSS, M.D. (1977a). Frequency-dependent fitness and differential outcrossing rates in hermaphrodite populations. Amer. Nat. 111: 200-202.

ROSS, M.D. (1977b). Behaviour of a sex-differential fertility gene in hermaphrodite populations. Heredity 38: 279-290.

ROSS, M.D. (1978). The evolution of gynodioecy and subdioecy. Evolution 32:174-188.

ROSS, M.D. (1982). Five evolutionary pathways to subdioecy. Amer. Nat. 119: 297-318.

ROSS, M.D. (1984a). Die Bedeutung der Sexualsysteme von Waldbaumarten. Forstarchiv 55:183-185.

ROSS, M.D. (1984b). Frequency-dependent selection in hermaphrodites: the rule rather than the exception. Biol. J. Linn. Soc. 23: 145-155.

ROSS, M.D. and H.-R. GREGORIUS (1983). Outcrossing and sex function in hermaphrodites: A resource-allocation model. Amer. Nat. 121: 204-222.

ROSS, M.D. and H.-R. GREGORIUS (1985). Selection with gene-cytoplasm interactions. II. Maintenance of gynodioecy. Genetics 109: 427-439.

ROSS, M.D. and B.S. WEIR. (1975). Maintenance of male sterility in plant populations. III. Mixed selfing and random mating. Heredity 35: 21-29.

ZIEHE, M. (1985). Polymorphic equilibria under inbreeding effects and selection on reproduction components. (This volume).

ZIEHE, M. and H.-R. GREGORIUS (1981). Deviations of genotypic structures from Hardy-Weinberg proportions under random mating and differential selection between the sexes. Genetics 98: 215-230.

PART III :

GENETIC DIFFERENTIATION WITHIN AND BETWEEN POPULATIONS

THE EFFECTS OF FOREST MANAGEMENT ON THE GENETIC VARIABILITY OF PLANT SPECIES IN THE HERB LAYER

W.H.O. Ernst

Abstract

The change of the genetic structure of three plant species of woodland clearings, *Senecio sylvaticus*, *Digitalis purpurea* and *Scrophularia nodosa* has been studied in relation to the management of these clearings.

Despite a lack of allozyme variation between and within populations of *Senecio sylvaticus* increase of the light accessibility to the forest floor by thinning has let evolve a winter annual type in addition to the common summer annual type. In contrast to the latter type the fruits of the former one germinated without cold treatment; seedlings developped a winter hardiness, but they took more time to flower and fruit than the summer annual type. In addition the winter annual type was physiologically differentiated in individuals with high and low chlorophyll content, the latter being correlated with a high magnesium and a low iron content of the leaves.

All populations of the biennial *Digitalis purpurea* consisted of plants with glabrous and hairy flower stalks, the latter being the recessive homozygote. Mowing of herbs and gresses, but not of shrubs and young trees in woodland clearings extended the survival of the populations for some more generations, and gave rise to a selection in favour of the glabrous genotype of *D. purpurea*.

Populations of the perennial *Scrophularia nodosa* contained iron efficient and inefficient plants. The ability to change the pH around the root seems to be coded by a single gene. In addition to this acidification Fe-efficiency was further determined by the Fe-translocation from the root to the shoot. Genotypes with a slow Fe-translocation were inferior in flower and seed production and unable to produce seeds if ploughing of woodland clearings before afforestation had diminished the soil organic matter of the A_o-horizon.

Introduction

After clear-felling, fire or storm damage regrowth of forest in the (sub)atlantic region of Europe is predominantly governed by annual, biennal and perennial herbs (Van Andel and Ernst, 1985); young tree seedlings contribute nearly no biomass to the new forest during the first five years. Generally, the pioneer species disappear within 3 to 6 years after clear-felling due to exhaustion of nutrients in the soil surface and/or to due their inferiority in the competition for water and light (Ernst and Nelissen, 1979; Van Baalen, 1982a). The pionier species of the herb layer will recolonize the same locality only after a complete tree generation, i.e. 30-200 years after their last colonies depending on the tree species. This recolonization may take place either by germination from a dormant seed bank with a high longevity, as in the case of *Digitalis purpurea* (Van Baalen, 1982b) or by the arrival of wind-dispersed diaspores as in the case of *Senecio sylvaticus*, *S. viscosus* and *Chamaenerion angustifolium*.

Every change in the forest management, e.g. ploughing the forest soil before replanting or mowing the herbs and gresses between young trees, but also thinning by man, acid rains or salt spray can influence the period of survival of the colonizing species. In this contribution I will demonstrate the effect of some of these changes on the genetic structure of these pionier species the annual pioneer species *Senecio sylvaticus*, the biennial *Digitalis purpurea* and the perennial *Scrophularia nodosa*. Due to the desparity of the variability of characters it is inevitable to report on different morphological and physiological characters for each species.

Material and Methods

All study sites are woodland clearings which have come into existence by tree-felling or by storm damage. In addition, on the Dutch coastal dunes populations of *Senecio sylvaticus* L. has been investigated in 40 to 60 years old, open Scotch and Austrian pine forests. The study sites for *Senecio sylvaticus* are those listed by Van Andel and Ernst (1985); those of *Digitalis purpurea* L. are given by Van Baalen (1982a), De Neeling (1982) and by Ernst and Vooijs (1985), those of *Scrophularia nodosa* L. by De Neeling (1982).

Allozyme analysis of 14 enzymes of all plant species and populations have been carried out according to the methods described by Koniuszek and Verkleij (1982). Crossing of *Senecio sylvaticus* was only possible by self-fertilization. The procedure of cross-pollination without emasculation, as described by Crisp and Jones (1978) and by Trow (1912) could not be used because a change in physiological characters is not *a priori* due to successful genetic crossings, as assumed by the above-mentioned authors for morphological traits. In the case of *Digitalis purpurea* and *Scrophularia nodosa* cross-pollination in the field as well in the greenhouse has been carried out after emasculation and protection of the treated flowers by plastic sacs, up to ten days after pollination.

Germination experiments have been carried out as described by Van der Vegte (1978), growth experiments according to Van Baalen (1982b) and Ernst and Vooijs (1985). Due to a strong interaction between initial germination behaviour of diaspores of *Senecio sylvaticus*, their position in the fruit stand and their final weight (Ernst, in prep.) only diaspores of the first five capitula of a plant have been used immediately after harvesting. To detect pH changes in the medium due to differences in iron demand of seedlings of *S. nodosa*, seeds of the original field populations and those of the F_1-plants were layed on 1.2% agar gel with a macronutrient solution and chlorophenol red as pH indicator, conform the method described by De Neeling (1982).

For the photoperiodic response of the winter and summer annual type of *S. sylvaticus* diaspores of the summer annual type were stratified for 3 months at 5°C and after germination transferred to the greenhouse. In each experiment 30 diaspores, of winter and summer annual type, which have emerged at the same day have been grown on a commercial soil (type Calceolaria) in a greenhouse at 75 $\pm$ 5% relative humidity, a day/night temperature regime of 20°/15°C and at two light periods 13 h or 17 h, supplied by mercury vapour lamps.

To investigate winter-hardiness of both types of *S. sylvaticus*, one week old seedlings (50 plants per type) were transplanted during a frost-free period in November in the open coastal pine forest and marked by plastic rings. Biweekly the survival of the plants has been registered. For the experiments testing the effect of fulvic and humic acid on the fertility of *S. nodosa* both organic soil components were extracted from a finehair podzol of the Veluwe as previously described by Marquenie-van der Werff and Out (1981).

Analysis of chlorophyll content of the leaves was the same as that described by Esashi et al. (1977), analysis of the concentration of mineral nutrients as that described by Ernst (1983a). Measuring of the transpiration rate was made conform the previously described procedure (Ernst et al. 1984).

Results and Discussion

1. *Senecio sylvaticus* - Differentiation in winter and summer annuals

Senecio sylvaticus, an autogamous plant with wind-dispersed diaspores is one of the pioneer species colonizing woodland clearings on acid to neutral soils already in the first year after clear-cutting, forest fire or storm damage (Van Andel and Ernst, 1985). It has no persistent seed bank, but it can persist in open forests if fresh litter will be applied regularly (Ernst and Nelissen, 1979).

Despite a permanent founder situation (Verkleij, 1980) there is a complete lack of allozyme variation (Koniuszek and Verkleij, 1982), which seems to be typical for inbreeding plant species (Hamrick et al., 1979). Even the extension of the allozyme analysis, based on 14 enzymes, to a population of *S. sylvaticus* in lower Austria, 1200 km south of the stands being analysed by Koniuszek and Verkleij (1982), did not give another result.

In contrast to the lack of within and between variability, based on allozyme studies, there is a high variability in quantitative characters. The lack of correlation between allozyme research and physiological as well as morphological characters is not astonishing because most physiological adaptation of plants to environmental factors are based on quantitative changes. Thus, if gene amplification or an increase in the copy number of the gene will be a common mechanism regulating these adaptations as demonstrated for copper resistance in yeast (Welch et al., 1983) then allozyme analysis will be not a meaningful tool for understanding plant population genetic.

Variation within a population will be demonstrated for a population, growing in the open coastal forest with *Pinus sylvestris* L. and *Pinus nigra* Arnold. In contrast to clearings *S. sylvaticus* is present in this coastal location at least for ten years, according to forest officers up to 40 years. These coastal populations consist of plants with the behaviour of either a winter annual or a summer annual (Van Andel and Ernst, 1985). The plants behaving as winter annuals germinate in the field in early autumn and flower in early summer of the next year, those behaving as summer annuals germinate in spring, flower in late summer and the diaspores remain dormant up to the folowing spring. From April onwards both annual types coexist in different phenological stages. The first obvious difference between summer and winter annual types is the germination behaviour of the diaspores. Similar as in summer and winter population of *Stellaria media* (Van der Vegte, 1978) freshly harvested achenes of the slowly growing winter generation shows a nearly complete germination at 20°C within 2 months (96.0±2.8%), whereas that of the summer annual type was only 8.0±3.6% after 2 months (Tab. 1). At low temperature (5°C) germination of the winter annual type was only 2.3±1.2% after 2 and 4 months, that of the summer annual type was 31.3±3.5%

time after imbibition (days)	summer annual type 5°C	20°C	winter annual type 5°C	20°C
30	8,3 ± 1,5	3,3 ± 3,1	0,0 ± 0,0	95,2 ± 2,1
60	31,3 ± 3,5	8,0 ± 3,6	2,3 ± 2,1	96,0 ± 2,8
90	76,0 ± 20,0	41,7 ± 4,7	2,3 ± 2,1	96,0 ± 2,8
120	95,7 ± 4,0	42,7 ± 4,0	2,3 ± 2,1	96,0 ± 2,8

Table 1. Percentage germination of freshly harvested achenes produced by a summer and a winter annual type of *Senecio sylvaticus* from an open Dutch coastal pine forest. Germination experiments with 3x100 seeds per set were carried out at constant temperature of 5°C and 20°C, respectively.

after 2 months, and 95.1±4.0% after 4 months. Independent of the germination temperature and the further growth conditions the winter annual type producesthe same quality of the achenes in the F_1 and F_2 generation.Due to the polymorphism of the seed germination behaviour of the and other summer annual types from Central Europe it can be assumed that the evolution of the winter annual type has its origin in a polymorphic summer annual type.

In addition to the difference in germination behaviour there is also a great difference in the growth rate between summer and winter annual type. Despite a general positive effect of the photoperiod on the mean development time from emergence to first flowering in annual plant species (Roberts and Feast, 1974) plants from the summer annual type took shorter to flower than plants of winter annual type, when emergence occurred at the same time (Tab. 2). Seedlings of the winter annual type, which germinated at 5°C, developped in the same manner as those which had germinated at 20°C. Both produce diaspores which behave as winter annuals. In contrast to the winter annual type, families, being derived from seedlings of a summer annual type which had germinated at 20°C as well as at 5°C contained plants which behaved either as summer or winter annuals, or as intermediates, suggesting that this trait of the summer annual type was polymorphic. In the Dutch coastal forest such intermediates flowering in late October instead of July, had been found at a density of 1 plant per ha, whereas the density of the winter annual type is about 15000 ha^{-1} and that of the summer annual type 3000 ha^{-1} plants. All plants had no vernalization requirements.

population	photoperiod (hours)	time from emergence to first flowering (days)	ratio 17/13 h photoperiod
Dutch coastal forest			
winter annual type	13	220 ± 26	
	17	92 ± 10	2,39
summer annual type	13	122 ± 6	
	17	74 ± 9	1,65
Teutoburg Forest/FRG	13	133 ± 8	
summer annual type	17	81 ± 12	1,64
Hunsrück Forest/FRG	13	124 ± 7	
summer annual type	17	78 ± 7	1,59
Ysper/Austria	13	96 ± 4	
summer annual type	17	61 ± 13	1,57

Table 2. Effect of photoperiod on the time from emergence to first flowering of plants from one winter and some summer annual types of *Senecio sylvaticus*.

With regard to winter-hardiness, which should be under polygenic control (Cahalan and Law, 1979), the winter annual type gave hundred percent survival of the plants, transplanted in early November to the open coastal pine forest, up to March and a 95% flowering response. The summer annual type was completely damaged by the frost in December in one experimental year and by frost in February in the other experimental year. This means that the survival of the summer annual type is only possible, if the plants germinate after the winter period.

In addition to the above mentioned differentiation within the coastal population of *S. sylvaticus*, plants of the winter annual type are also differentiated in their chlorophyll content (Tab. 3). Plants with pale green leaves, varying from 8 to 14.8% of the population within 5 years, have less than half as much chlorophyll as plants with normal green leaves; the chlorophyll content of the latter ones is the same as in the summer annual type of the coastal population and the same as that in all Central European populations, ranging from 1.04 mg up to 1.43 mg chlorophyll g^{-1} fresh weight. The ratio chlorophyll a/b is in all plants nearly the same, 3,32±0,22 in normal green leaves, 3,12±0,36 in pale green leaves. This difference in chlorophyll content remains constant in the F_1-inbreds of both types, perhaps due to cytoplasmic effects as found in other plant species (Mehndiratta and Phul 1983; McCashin and Canvin, 1979).

leaf colour		chlorophyll content (mg g^{-1} fresh weight)	concentration (µmol g^{-1} dry weight) of Fe	Mg	K	Na
normal	green					
	field	1.23±0.15	14.6±2.5	93±13	411±213	369±174
	F_{1a}	1.14±0.10	9.1±2.6	113± 3	603± 22	282± 31
	F_{1b}	1.17±0.12	13.9±3.5	97± 5	245± 26	486±218
pale	green					
	field	0.48±0.06	6.1±1.3	175± 9	409± 59	636±145
	F_1	0.43±0.16	2.3±0.9	177± 8	376± 8	734± 70

Table 3. Chlorophyll content and concentration of some mineral nutrients in the leaves from normal and pale green plants of the winter annual type of *Senecio sylvaticus*. Five plants of every group have been selected in the field, their F_1-progeny has been grown in a greenhouse at a light/dark regime of 17/7 hours, a day/night temperature regime of 20°/15°C and at 75±5% relative humidity.

In contrast to short-living populations of *S. sylvaticus* on woodland clearings the coastal population families of the pale and normal green plants are also differentiated with regard to their uptake of mineral nutrients (Tab. 3). The most obvious is the high iron concentration and the low magnesium concentration of normal green plants and the low iron and high magnesium concentration in the pale green plants. Due to a lack of information on the mineral status of plants with different genetically based chlorophyll contents (Turcotte and Feaster, 1973; Moret-Gaudry et al., 1978; Secor et al., 1982) and the recent findings of the physiology of iron efficiency and inefficiency (Bienfait and Van der Mark, 1983; Cline et al., 1984; see also in this contribution *Scrophularia nodosa*) in this context, the statement may be sufficient that some decades are necessary to let differentiate a 'permanent' founder with low genetic variability to a highly physiologically variable permanent population.

2. *Digitalis purpurea* - Hairiness of the flowering axis

Digitalis purpurea, a short-living perennial with a very persistent seed bank (Van Baalen, 1982b), flowers in the second summer after germination and can flower once more in the third summer under exceptional conditions (Van Baalen and Prins, 1983). Aside from a flower colour polymorphism which is scarce in most field populations (Ernst and Vooijs, 1985) *D. purpurea* exhibits a distinct variation of the hairiness of the lower part of the flowering axis, described as the forma *pubescens* and the forma *nudicaulis*, both determined by two alleles of one locus, forma *pubescens* being the recessive homozygote (Saunders, 1918). As demonstrated in table 4 and 5, all investigated populations of *D. purpurea* contain both forms. The gene frequency of

population	altitude (m above sea level)	number of counted plants	number of hairy plants	frequencie of the recessive gene
The Netherlands:				
Elzet/Limburg	200	994	855	0.9274
Bovenste Bos/Limburg	210	294	152	0.7190
Belgium:				
Louveigne	270	370	340	0.9586
Germany:				
Lautenthal/Hartz	480	189	135	0.8452
Wildemann/Hartz	500	528	482	0.9554
Iberg/Hartz	510	57	56	0.9912
Hahnenklee/Hartz	560	322	270	0.9157
Harzburg/Hartz	650	75	64	0.9238
Iburg/Teutoburg Forest	250	252	244	0.9840
Blankenrode/Sauerland	360	117	102	0.9337
Bundenbach/Hunsruck	390	130	120	0.9608
Schnepfenbach/Hunsruck	420	213	202	0.9738
Morsbach/Hunsruck	650	485	411	0.9206
Spain:				
Santiago de Compostela	180	97	45	0.6811

Table 4. The frequency of the recessive homozygote (hairy stem) in some European populations of *Digitalis purpurea*. Data for the populations in the Netherlands and in Belgium are based on counts by Van Baalen (1982a), the other data are based on counts by Ernst (unpublished).

the recessive homozygote varied from 0.0969 up to 0.9912; only in one population it was below 0.5.

The advantage of the recessive homozygots is not due to a difference in germination behaviour, as it was found for the seeds of various flower colour morphs of *D. purpurea* (Ernst and Vooijs, 1985). An analysis of the gene frequency from two subsequent generations, i.e. two years after flowering and seed shedding from the former generation, demonstrated that in all populations investigated by Van Baalen (1982a) during woodland succession, there was a decrease of the percentage of the recessive genotype independent of the decreasing or increasing number of individuals (Tab. 5). The change in gene-frequency varied from 0.4% up to 5.2%; in four of the six populations the difference was less than 1%.

One reason for the change in gene-frequency may be the change in some environmental conditions, especially the water balance, during the vegetation succession from a clearing to a forest (Bjor, 1972). With the development of a shrub layer (*Betula alba* and *Quercus robur*) the air humidity near the soil surface increased and was higher than in the open clearing. The forma *nudicaulis*, which had not only a smooth stem, but also less hairs on the leaf surface, had a more than 20% increased transipration rate in comparison to the hairy plants (68±9 mg H_2O $g^{-1}h^{-1}$ vs. 55±5 mg H_2O g^{-1} fresh weight h^{-1}). Therefore the more efficient water economy of the hairy form may favour this genotype in an open clearing, but not in a closed herb and semi-closed shrub layer.

population	number of counted plants	number of hairy plants	frequency of the recessive gene(q)	$\Delta q = q_1 - q_0$	selection coefficient s
Mechelen	90	2	0.1491		
1979	213	2	0.0969	-0.0522	1.561
Ter Graat					
1979	35	31	0.9411		
	29	23	0.8906	-0.0505	0.523
Vaals II					
1979	80	44	0.7416		
1981	205	110	0.7325	-0.0091	0.09
Elzet I					
1979	275	237	0.9283		
1981	98	83	0.9203	-0.0080	0.116
Elzet II					
1979	150	128	0.9238		
1981	120	101	0.9174	-0.0064	0.091
Vaals I					
1979	290	255	0.9377		
1981	474	412	0.9323	-0.0054	0.091

Table 5. The change of the gene frequency of the recessive homozygote (hairy form) in some Dutch populations of *Digitalis purpurea* after one generation (two years interval). Data are based on the counts by Van Baalen (1982a). The altitude of the stands varies between 160 and 300 m above sea level.

The change of gene-frequency of *D. purpurea* on a woodland clearing (Vaals I) where since 1981 every year after seed shedding of *D. purpurea* all herbs and grasses, but not the birch and oak shrubs have been mown, support this interpretation. In contrast to untreated clearings with only two or three generations of *D. purpurea*, this population produced already 5 generations. In 1984 the gene frequency of the recessive homozygote was 0.8175 (422 plants being counted). Within three generations, the selection coefficient has increased from 0.091 to 0.5581 i.e. selection was strongly against the recessive homozygote.

3. *Scrophularia nodosa* - Iron efficiency

The perennial plant species *Scrophularia nodosa* is a rhizome geophyte occurring in early as well in later stages of succession in woodland clearings. Despite simultaneous flowering of this entomogamous species Koniuszek (Thesis, in preparation) established only a low selfing frequency on the basis of allozyme studies. The investigated populations have a high degree of allozyme polymorphism (Koniuszek, in prep.). They are also variable with regard to their iron metabolism; every population contains individuals which can be determined as Fe-efficient and Fe-inefficient according the terminology of Brown and Ambler (1974). Due to a new technique for measuring iron efficiency already in the seedlings stage (De Neeling, 1982) it was possible to analyse the degree of Fe-efficiency on the population level. In all populations, independent of their occurrence on alkaline or acid soils, 70 to 85 per cent of the plants have the ability to acidify the medium in the vicinity of the roots (Van Baalen et al., 1984). This result is in good agreement with those found in populations of *Chamaenerion angustifolium* (De Neeling, 1982) and *Senecio fuchsii* (Ernst, 1983b). The inheritance of this ability seems to be coded by a single gene. All F_1 plants from reciprocal crosses changed the external pH, independently of which parent was used as female. The F_2 progeny segregated in a 3:1 ratio (efficient:inefficient), being in agreement with the results in soybeans (Weiss, 1943).

After transplantation of the seedlings of *S. nodosa* to a commercial soil a further differentiation with regard to iron uptake and translocation in the plant has been found. Plants can be distinguished by the accumulation pattern of iron in young and old leaves. Twenty per cent of the efficient plants accumulate high amounts of iron in old leaves and only small amounts in young ones, defined as the 'slow translocation type' (Ernst, 1983b). In nearly 60 per cent of the plants iron has been evenly distributed on old and young leaves (Tab. 6). A further 20 per cent of the plants accumulate more iron in young than in old leaves (fast translocation type).

Up to now, the analysis of the genetic basis of the translocation types is more complicated because a lot of crosses produced either poor seeds or seeds with a poor germination. In every case, iron efficient plants do not only possess a single gene

seedling	translocation	iron concentration in young leaves	old leaves	significance with regard to leaf age
inefficient	equal	1.01±0.13	1.03±0.97	P<0.05
efficient	slow	1.14±0.48	3.41±0.26	P<0.001
efficient	equal	1.35±0.40	1.59±0.39	not significant
efficient	fast	3.07±0.85	1.70±0.31	P<0.05

Table 6. Iron concentration (µmol g^{-1} dry weight) of young and old leaves of *Scrophularia nodosa* plants at the time of flowering in relation to the ability of the seedlings to change the pH in the vicinity of the roots (inefficient plant cannot, efficient plant do change the pH).

for acidification of the medium around the root, but also one or more genes being responsible for the iron translocation from root to shoot. The ecological consequences of the different translocation types becomes clear if the total demand of the plants is considered (Tab. 7). The demand is high in the slow translocation

translocation type	amount of iron (µmol per plant)
slow	42.6 ± 7.4
equal	10.2 ± 2.0
fast	5.0 ± 1.5

Table 7. Amount of iron in the total plant (root, shoot, leaves) of plants of *Scrophularia nodosa* with different iron translocation behaviour. All values are significantly different at P < 0.001.

type and low in the fast translocation type. Due to the high demand of iron for the production of flowers (0.34±0.20 µmol g^{-1} dry flower) and capsules (2.23±0.50 µmol g^{-1} dry weight) plants with a slow iron translocation produce less flowers (56±15 versus 207±59) and seeds (up to 90%) than plants belonging to the equal and fast translocation type. This effect can be increased by the interaction with such soil organic components which can strongly complex iron (Van der Werff, 1981). If ploughing of woodland clearings will precede planting of young trees, this way of soil management has not only a tremendous negative effect on plant growth (Ernst and Nelissen, 1979), but especially on the generative reproduction of herbs. In the case of *Scrophularia nodosa* iron efficient plants with equal translocation produce only 126 ± 66 µg flowers per plant in the absence of fulvic and humic acids (B_3-horizon of the soil profile) whereas in the presence of both soil organic acids in a ratio from 1:2 up to 2:1, being typical for A_0/A_1-horizon of brown earth, the flower production achieve 214 ± 32 µg per plant, i.e. nearly 70% more flowers and later more seeds. In the absence of organic matter the Fe-efficient plants with the slow iron translocation produce no seeds. This means that soil management as the ploughing of clearings before afforestation can influence the genetic structure of populations of *S. nodosa*.

References

Bienfait, H.F. and F. van der Mark. (1983). Phytoferritin and its role in iron metabolism. Metals and Micronutrients. Uptake and Utilization by Plants, edited by D.A. Robb and W.S. Pierpoint, pp. 111-123. London: Academic Press.

Bjor, K. (1972). Micro-temperature profiles in the vegetation and soil surface layers on uncovered and twig covered plots. *Medd. Norske Skogforsoksvesen* 30: 203-218.

Brown, J.C. and J.E. Ambler. (1974). Iron-stress response in tomato (*Lycopersicon esculentum*). I. Sites of Fe reduction, absorption and transport. *Physiol. Plant.* 31: 221-224.

Cahalan, C. and C.N. Law. (1979). The genetical control of cold resistance and vernalisation requirement in wheat. *Heredity* 42: 125-132.

Cline, G.R., C.P.P. Reid, P.E. Powell and P.J. Szaniszlo. Effects of a hydroxamate siderophore on iron absorption by sunflower and sorghum. *Plant Physiol.* 76: 36-39.

Crisp, P. and B.M.G. Jones. (1978). Hybridization of *Senecio squalidus* and *S. viscosus* and introgression of genes from diploid into tetraploid *Senecio* species. *Ann. Bot.* 42: 937-944.

De Neeling, A.J. (1982). Adaptation of plants from clearings to acid and alkaline soil. Ph.D. Thesis, Free University Amsterdam.

Ernst, W.H.O. (1983a). Element nutrition of two contrasted dune annuals. *J. Ecol.* 71: 197-209.

Ernst, W.H.O. (1983b). Ökologische Anpassungsstrategien an Bodenfaktoren. *Ber. Deutsch. Bot. Ges.* 96: 49-71.

Ernst, W.H.O. and H.J.M. Nelissen. (1979). Growth and mineral nutrition of plant species from clearings on different horizons of an iron-humus podsol profile. *Oecologia* 41: 175-182.

Ernst, W.H.O., A.J. De Neeling and H. Vooijs. (1984). Replacement of sodium and potassium in a natrophobe and a natrophile species. *Z. Pflanzenphysiol.* 122: 147-154.

Ernst, W.H.O. and R. Vooijs. (1985). Scarcety of flower colour polymorphism in field populations of *Digitalis purpurea*. *Oecologia* (submitted).

Esashi, Y., Y. Katoh, Y. Hata and N. Goto. (1977). Dormancy and importency of cocklebur seeds. 7. Inability of dormant cotyledons to form chlorophyl. *Plant Physiol.* 59: 122-125.

Hamrick, J.L., Y.B. Linhart and J.B. Mitton. (1979). Relationships between life history characteristics and electrophoreticlly detectable genetic variation in plants. *Ann. Rev. Ecol. Syst.* 10: 173-200.

Koniuszek, J.W.J. and J.A.C. Verkleij. (1982). Genetical variation in two related annual *Senecio* species occurring in the same habitat. *Genetica* 59: 133-137.

Marquenie van der Werff, M. and T. Out. (1981). The effect of humic acid as Zn complexing agent on water culture of *Holcus lanatus*. *Biochem. Physiol. Pflanz.* 176: 274-282.

McCashin, B.G. and D.T. Canvin. (1979). Photosynthetic and photorespiratory characteristics of mutants of *Hordeum vulgare*. *Plant Physiol.* 64: 354-360.

Mehndiratta, P.D. and P.S. Phul. (1983). Genetic and cytoplasmic effects on chlorophyll content in pearl millet (*Pennisetum typhoides*). *Theor. Appl. Genet.* 65: 339-342.

Moret-Gaudry, J.F., D.A. Thomas, M.E. Deroche and P. Chartier. (1978). Growth, leaf optical properties, chlorophyll content and net assimilation rate in maize seedlings with and without the gene Opaque 2. *Photosynthetica* 12: 284-289.

Roberts,H.A. and P.M. Feast. (1974). Observations on the time of flowering in mayweeds. *J. Appl. Ecol.* 11: 223-229.

Saunders, E.R. (1918). On the occurrence, behaviour and origin of a smooth-stemmed form of the common foxglove (*Digitalis purpurea*). *J. Genetics* 7: 215-228.

Secor, J., D.R. McCarty, R. Shibles and D.E. Green. (1982). Variability and selection for leaf photosynthesis in advanced generations of soybeans (*Glycine max*). *Crop Sci.* 22: 255-259.

Trow, A.H. (1912). On the inheritance of certain characters in the common groundsel - *Senecio vulgaris*, Linn. - and its segregates. *J. Genetics* 2: 239-277.

Turcotte, E.L. and C.V. Feaster. (1973). The interaction of 2 genes for yellow foliage in cotton. *J. Hered.* 64: 231-232.

Van Andel, J. and W.H.O. Ernst. (1985). Ecophysiological adaptation, plastic responses, and genetic variation of annuals, biennials and perennials in woodland clearings. Structure and Functioning of Plant Populations II: Phenotypic and Genotypic Variation in Plant Populations, edited by J. Haeck and J.W. Woldendorp. Amsterdam; Verh. Kon. Ned. Akad. Wetensch. 85 (in press).

Van Baalen, J. (1982a). Population biology of plants in woodland clearings.Ph.D. Thesis, Free University, Amsterdam.

Van Baalen, J. (1982b). Germination ecology and seed population dynamics of *Digitalis purpurea*. *Oecologia* 53: 61-67.

Van Baalen, J. and E.G.M. Prins (1983). Growth and reproduction of *Digitalis purpurea* in different stages of succession. *Oecologia* 58: 84-91.

Van Baalen, J., H.J.M. Nelissen, W.H.O. Ernst, J. Wattel and R. Vooijs. (1984). Reproductive processes in *Senecio fuchsii* (partly in comparison with *Eupatorium cannabinum*) as affected by temperature, irradiance and soil fertility. *Flora* 175: 81-90.

Van der Vegte, F.W. (1978). Population differentiation and germination ecoloyg in *Stellaria media* (L.) Vill. *Oecologia* 37: 231-245.

Van der Werff, M. (1981). Ecotoxicity of heavy metals in aquatic and terrestrial higher plants. Ph.D. Thesis, Free University, Amsterdam.

Verkleij, J.A.C. (1980). Isoenzyme variation in pioneer species from ecologically different habitats. Ph.D. Thesis, Free University, Amsterdam.

Weiss, M.G. (1943). Inheretance and physiology of efficiency in iron utilization in soybeans. *Genetics* 28: 253-268.

Welch, J.W., S. Fogel, G. Cathala and M. Karin. (1983).Industrial yeasts display tandem gene iteration at the CUP1 region. *Mol. Cell. Biol.* 3: 1353-1361.

STUDIES ON BREEDING STRUCTURE IN TWO TROPICAL TREE SPECIES*

Kan-Ichi Sakai

ABSTRACT

Two tropical forest tree species, *Altingia excelsa* Noronha and *Agathis borneensis* Warb. were investigated for their breeding structure in natural forests. Assuming that inbreeding would produce higher similarity than random mating in the pattern of isoperoxidase bands in individual trees, the degree of disagreement between trees in isoperoxidase patterns was calculated for estimating occurrence of inbreeding. By finding that trees alike in the isoperoxidase patterns were forming separate groups in the forest in *Altingia*, five family clumps were presumed(Figure 3). It was found that those different families were dissimilar in respect of isoperoxidase constitution as well as in several leaf characters. The distance between two trees at which they could mate was estimated to be 16 to 18 meters and the area one family occupies 200 to 250 m² on the assumption that inbreeding in dioecious *Altingia* should have occurred by consanguineous mating within each family. In spite of growing together of two families within the mating distance, it was found that they showed apparent genetic differentiation between them suggesting that they had been sexually isolated from each other. This sexual isolation, if it happened, is supposed to be due to genetic difference in flowering time.

In *Agathis*, there was no indication of family clump formation. Thus, the inbreeding effect observed in this monoecious species is supposed to have come from self-fertilization of individual trees.

---

* This investigation was conducted in BIOTROP(SEAMEO) in Bogor, Indonesia, under the international cooperation of Japan and Indonesia. The scientists taking part in this study were Drs. Toru Endo, Shinya Iyama, Yasusada Miyazaki, Shigesuke Hayashi, Yoshiya Shimamoto, Lilian U. Gadrinab and Ulfah Juniarti besides the present author. The full particaulars with citations will be published in BIOTROP BULLETIN in future.

INTRODUCTION

The tropical rain forest stimulates our interest in enquiring into the breeding structure of trees, because it is widely believed that species diversity is high in tropical forests which necessarily brings about low density of trees of each species. Do trees propagate by inbreeding under such a low density? Are trees of the tropical species free from or resistant to inbreeding depression? To what extent are they genetically differentiated within a species?

In this paper the writer intends to report results of the study with two species, Altingia excelsa Noronha (Hamamelidaceae) and Agathis borneensis Warb. (Araucariaceae) growing in natural forests in West Java and Kalimantan, respectively. This study aims at estimation of degree of inbreeding occurring in nature and to know how those trees which are supposed to be consanguineous distribute and propagate within a stand, having in mind an object to find a way to pierce an opening to the problem of speciation in the tropical rain forest.

MATERIALS AND METHODS

The stand in which Altingia excelsa trees were investigated was located on the outskirts of the village of Ciwidey near Bandung in the Western Java. Number of trees of various species with DBH exceeding 6 cm growing in a quadrat of 50 x 50 m in the stand was 148 in all, among which 38 were trees of Altingia excelsa. The 38 Altingia trees were measured for their growth, and their position in the quadrat was mapped on a section-paper. Twigs with mature leaves were collected from each of them and were divided into two parts, one for measurement of leaf characters while the remaining part for the electrophoretic analysis of the isoperoxidases. Two other enzymes, i.e. acid phosphatase and esterase, were also examined, but they were rather simple in pattern of the zymograph and not used in this study.

Agathis borneensis was sampled from a quadrat of the same size plotted in a natural forest of International Timber Cooperation Indonesia in Kalimantan. There were about 230 trees with DBH exceeding 6 cm of various species in the quadrat, of which 21 were trees of Agathis borneensis. Growth measurement and mapping of individual trees were made and their leaves on an individual tree basis were collcted for the study of leaf characters and electrophoretic analysis.

Leaf characters measured in both species were leaf length, leaf width, leaf size(length x width), number of veins, and length and thickness of petioles.

The peroxidase zymograms were obtained as follows: One hundred mg of finely cut leaf pieces were crushed in a mixture of 100 mg of quartz sand, 50 mg of PVPP(polyvinyl-polypyrrollidone) and 0.5 ml of the enzyme extractant, pH 7.5, containing 10 % Triton X-100, 0.5 M NaCl and 0.2 M ascorbic acid. The gel buffer contains 0.42 g of histidine HCl and 0.072 g NaOH per 0.5 l, being mixed with 60 g of hydrolyzed starch. The tray buffer, pH 6.0, contains 117.6 g of trisodium citrate and 9.4 g of citric acid per litre. The leaf extracts were run under 250 V per 20 cm for 4.5 hours. The zymograms were assayed in a mixture of 0.05 M Tris acetate buffer, pH 4.0, 0.03 % H_2O_2, 4 mM eugenol, 4 mM 3-amino-9-ethylcarbazole and 10 % acetone.

On an assumption that the pattern of the isoperoxidase bands of a tree would be likely to reflect its genetic back-ground, genotypic dissimilarity among trees was estimated by measuring the degree of between-tree disagreement in their isozyme patterns. It has been called as the disagreement count and is taken to be indicative of genotypic dissimilarity between trees. In practice, there may occur variation in activity of bands either genetically or non-genetically, but it is neglected for the present study, only taking into consideration the presence or absence of bands confronting each other. The method of calculation is very simple as given in Figure 1.

Detection of occurrence of inbreeding and formation of family clumps has been made by the aid of the disagreement counts in the following way: It is assumed that plural isozyme bands would be distributed at random among individual trees if mating occurs at random, though the effect of linkage disequilibrium, if any, would arrest the random distribution to some extent. Inbreeding, if it occurs, may promote similarity in the isozyme pattern among sib-trees and at the same time dissimilarity

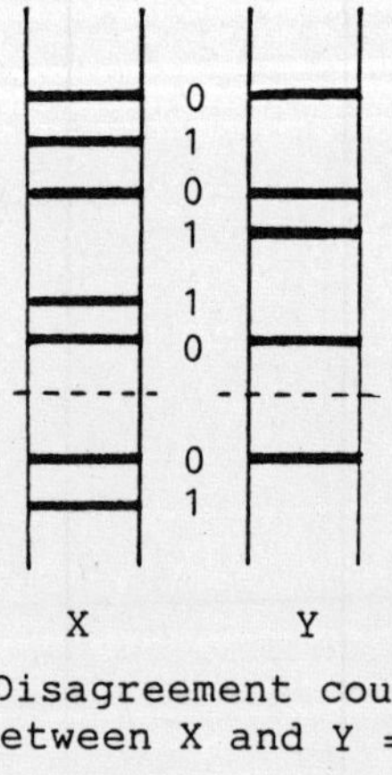

Figure 1. Schematic explanation for measuring the disagreement count between X and Y.

among non-related trees. Thus, disagreement counts obtained among trees of the same family are likely to be low in value. It should be borne in mind, however, that converses are not always true; Trees with low disagreement count cannot immediately be regarded as sib-members of a family. The evidence needed for identification of consanguinity among individual trees in a population would be that trees connected with low disagreement counts are distributed not at random, but are assembled in close proximity in groups, because saplings can generally be more likely to grow near the mother tree than at random in the forest. Consanguineous group would also be shown by inter-group differentiation in genetic characters including isozyme constitution and some vegetative characters.

RESULTS OF STUDY

I. Altingia excelsa Noronha

(A) Occurrence of inbreeding and formation of family clumps

Individual variation of DBH of 38 trees in the natural stand of Ciwidey is presented in Table 1. It shows that trees exceeding 61 cm in DBH were five among 38, or 13 %.

Table 1. Frequency distribution of DBH in 38 trees of Altingia excelsa growing in a 50 x 50 m quadrat.

	D B H in c m					Total
	6-20	21-40	41-60	61-80	81-100	
No. of trees	5	12	16	3	2	38
Frequency, %	13.16	73.68		13.16		100

In order to examine if the disagreement counts among 38 trees distributed in accordance with the random combination of isoperoxidases, the actual distribution was compared with the expectation calculated on the basis of random combination of 16 isoperoxidases. The result is presented in Table 2.

Table 2. Comparison of the actual observations with the theoretical expectations of disagreement counts in Altingia.

	Disagreement count						Total
	0,1	2	3	4	5	6 ≦	
Expectation	18.7	54.6	110.7	152.8	152.1	214.1	
Observation	51	70	95	116	117	254	703

$\chi^2 = 86.76$ d.f. = 5 $P < 0.01$

Data presented in Table 2 as well as in the dotted parts between two curves in Figure 2 show apparently the excess occurrence of observations over expectations in the lower classes and also in the higher classes.

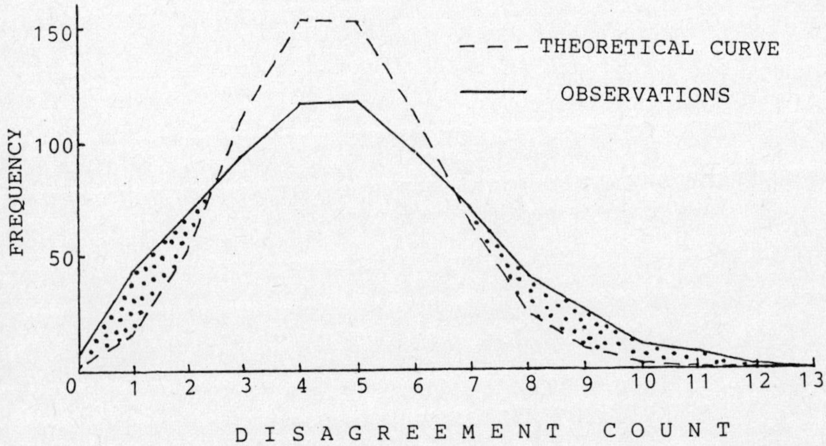

Figure 2. Comparison between observed and expected frequencies of disagreement counts in Altingia excelsa.

This is taken as to indicate that mating in Altingia is not at random, but a fair amount of inbreeding should be occurring, by which isozyme pattern among some individual trees are more alike and among some others are less alike than the random assortment of isozyme bands.

Inbreeding in a natural stand may occur in two ways: self-fertilization of individual trees and mating among relatives. Since Altingia is dioecious, the former is not expected to occur. For the latter to occur trees of the same family should stand in neighborhood, and/or they come to anthesis at the same time, but not synchronous with others. What is the actual situation in Altingia excelsa?

Table 3 shows the result of a test if tree pairs with smaller disagreement counts grow in the neighborhood. It is found from the table that in the box of 0,1 disagreement counts x 0-10 m intertree distance, a conspicuous excess of observation over expectation based on random distribution was detected, indicating that trees with similar isozyme patterns tended to grow collectively in the neighborhood. In other words, in Altingia excelsa genotypically alike trees are likely to grow together in groups in a natural forest. Thus, five groups in the forest have provisionally been defined as A to E family clumps(Figure 3). Broken lines connecting individual trees with arabic numerals 0, 1 or 2 in the

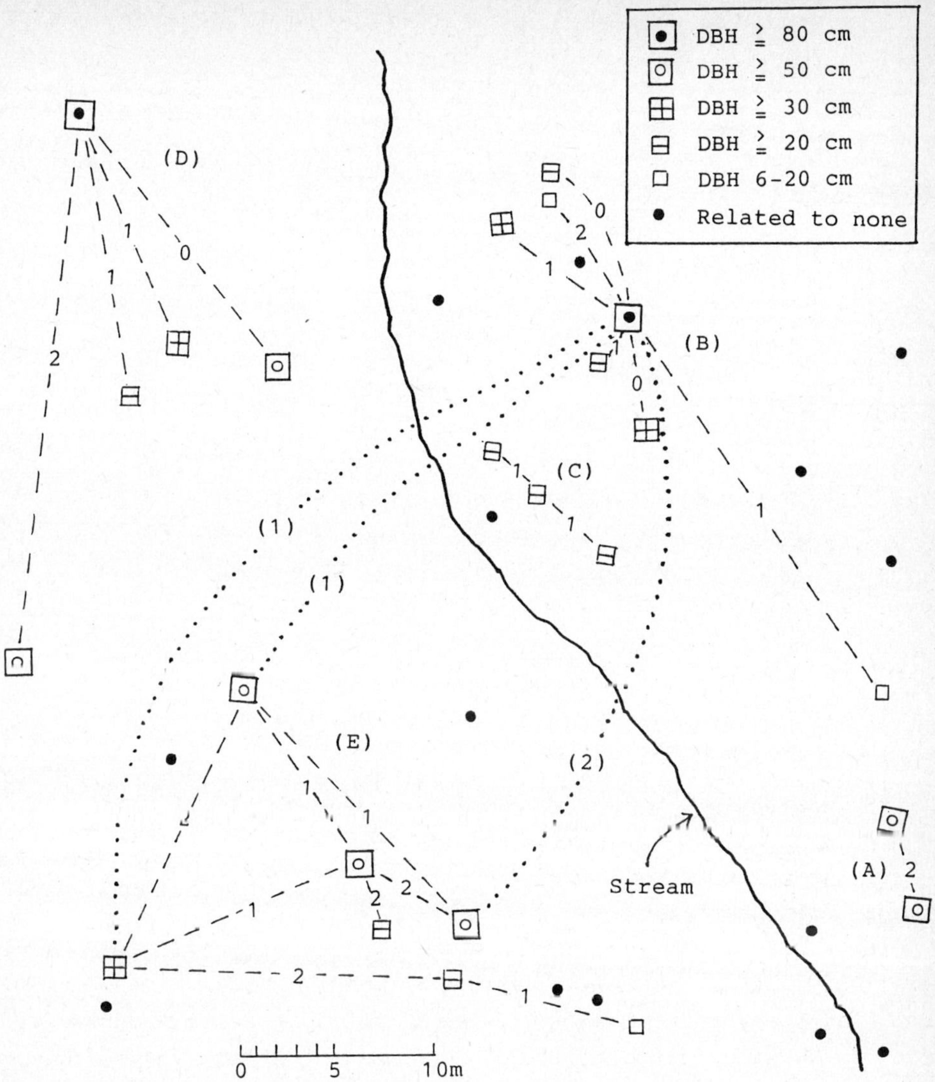

Figure 3. Family clumps in Altingia excelsa. Arabic numerals represent disagrement counts, while those in parentheses disagreemnt counts between trees of presumed relation.

Table 3. Relationship between disagreement counts and intertree distances in <u>Altingia excelsa</u>.

Intertree distance (m)		Disagreement count				Total	χ^2	P
		0,1	2,3	4,5	6 $\leqq$			
0-10	Exp.*	6.7	21.6	30.5	33.3			
	Obs.*	15	25	25	27	92	13.17	< 0.01
11-20	Exp.	12.3	39.8	56.3	61.6			
	Obs.	11	29	70	60	170	6.43	> 0.05
21-30	Exp.	14.3	46.4	65.5	71.7			
	Obs.	15	42	58	83	198	3.05	> 0.20
31 $\leqq$	Exp.	17.7	57.2	80.8	88.4			
	Obs.	10	69	80	85	244	5.90	> 0.10

* Expectation and observation.

figure are presumed ties between parent and child trees, the numerals speaking for the corresponding disagreement count. Of five groups in the figure, two, i.e. B and E are worthy of note because they are both largest groups each with 7 trees, and the largest trees of both groups are connected with small disagreement count 1 or 2, suggesting that these two groups might have been descendants of a common progenitor. Of more interest is that both groups appeared to show a within-group segregation in width of leaves (see Table 6). Although we are not yet in a position to speak definitely, it is suspected that B and E groups might be both heterozygous for a gene probably with major effect on leaf width.

(B) <u>Intergroup differentiation in the constitution of isoperoxidase components</u>

Occurrence of various isoperoxidase bands in trees of five groups is presented in Table 4.

Table 4. Interpopulation differentiation in distribution of isoperoxidase bands

Group	Number of trees	Isoperoxidase band (%)*							
		a	c	d	e	f	g	k	m
A	2	100	50	100		100		100	
B	7	100				100			100
C	3				100		100	100	100
D	5	100	100			100	60		100
E	7	43	43			100			86

* The maximum number of isoperoxidase bands per tree in <u>Altingia</u> <u>excelsa</u> was 16.

It is found from Table 4 that trees of each group are characterized by possessing specific isoperoxidase bands. For instance, C group is very peculiar in not having such bands as "a" and "f" which seem to be rather ubiquitous in the present species. Instead, it possesses "e", "g" and "k" bands which are more or less uncommon to other groups. Table 4 thus tells us that five groups are genetically differentiated in biochemical characteristics. It is interesting to find from the table that B and E groups are rather similar in having "a", "f" and "m" bands, though they differ in "c".

(C) Inter-population differentiation in leaf characters

Now we shall further enquire into differentiation among groups in some leaf characters. Results of analysis of variance are given in Table 5 which show that families are significantly variable in five characters except for the vein number.

Table 5. Analysis of variance of some leaf characters of 5 family groups of *Altingia excelsa*.

Source	d.f.	Mean squares					
		Leaf length	Leaf width	Leaf size	Vein number	Petiole length	Petiole thickness
Between families	4	7179*	1708**	9407**	10.5$^{n.s.}$	1186**	6073*
Within families	19	2039	307	1875	13.2	86	1988

*, ** Significant at the 5% and 1% levels, respectively.
n.s. Non-significant

The actual variation in six leaf characters is shown in Table 6. From the mean values described in the righthand column of the table, we find that B family had the biggest size in most characters, E the next, while C and D families were the smallest, leaving A intermediate.

From several facts given above, we would probably make no mistake in saying that tree groups A to E which have been defined up to now as presumed families on the basis of isoenzyme patterns are actually clumps of tres of individual families or family clumps each of which being genetically differentiated from others within the forest.

For genetical differentiation among family clumps to occur, it is necessary that propagation occurs by mating within the same, but not between different, families. Now, let us assume that trees of a single family form a breeding group within which mating occurs at random, then what would be the area occupied by a family or the size of the so-called neighborhood. Of five families, A and C are discarded from the following enquiry because of a too small number of trees in them.

The inter-tree distances in three families, B, D and E, were measured (Table 7), and their distribution within each family is shown in

Figure 4. In Table 7 and Figure 4, measurement was made in two ways: One with all trees belonging to each family, while another for all trees but one standing apart, because in Figure 3, we notice that in each of three families, one tree is always standing relatively far apart from main constituents of the same family. The distance between the farthest tree and the remaining ones are depicted in black circles in Figure 4.

Table 6. Variation of six leaf characters in five family groups of <u>Altingia excelsa</u>

Character	Family	Number of trees	Class value*							Mean
			1	2	3	4	5	6	7	
Leaf length (mm)	A	2	1				1			115.8
	B	7			1	2	2	1	1	134.9
	C	3	1	2						104.4
	D	5	1	2	2					109.8
	E	7	1	2	1	2		1		117.8
Leaf width (mm)	A	2		1			1			50.7
	B	7				2		5		60.9
	C	3		1	1	1				48.1
	D	5		2	3					47.2
	E	7			4			3		53.7
Leaf size (mm²)	A	2	1				1			60.0
	B	7				2	2	2	1	83.1
	C	3	1	2						50.8
	D	5	1	2	2					52.4
	E	7	1	2	1	1	1	1		64.3
Number of leaf veins	A	2		1	1					13.0
	B	7		5	2					12.5
	C	3		1	2					12.9
	D	5		2	2	1				13.3
	E	7		2	4	1				13.5
Petiole length (mm)	A	2					1	1		36.0
	B	7			2	5				27.4
	C	3		3						18.1
	D	5		2	2	1				22.5
	E	7		3	3	1				23.1
Petiole thickness (1/100 mm)	A	2		1	1					106.5
	B	7			1	2	1	2	1	130.9
	C	3		3						103.3
	D	5	1	1	2			1		110.9
	E	7	1	1	2	1	2			112.4

* One unit of class value corresponds to 10 mm for leaf length, 5 mm for leaf width and petiole length, 10/100 mm for petiole thickness, 10 mm² for leaf size and 2 for vein number.

Table 7. Average intertree distances in five families of *Altingia excelsa*

Family	All trees are considered				The farthest tree is excluded			
	Number of		Distances (m)		Number of		Distances (m)	
	trees	I.D.*	Mean	SD	trees	I.D.*	Mean	SD
A	2	1	4.14	-	-	-	-	-
B	7	21	12.42	9.56	6	15	7.09	3.71
C	3	2	3.50	-	-	-	-	-
D	5	10	13.97	7.08	4	6	10.10	5.27
E	7	21	12.57	6.61	6	15	10.79	5.71
Mean**	6.33	17.33	12.99	7.75	5.33	12.00	9.33	4.90

* Number of intertree distances measured.
** Mean of 3 families: B, D and E.

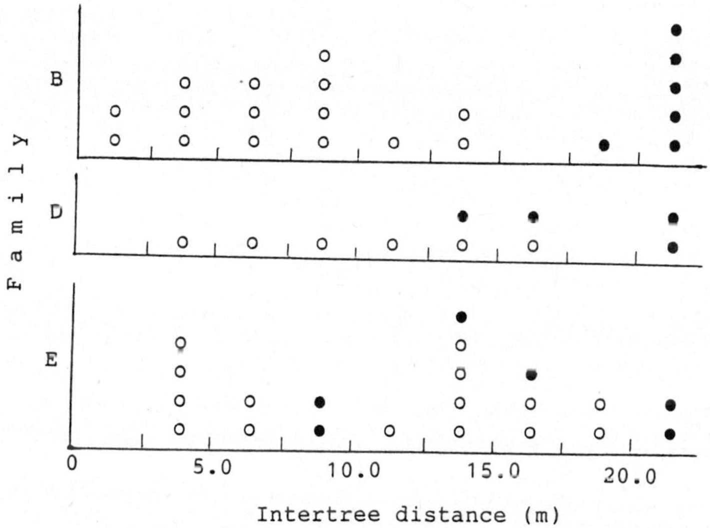

Figure 4. Distribution of intertree distances in three families of *Altingia excelsa*. Black circles are those between the farthest tree and the remaining ones in each family.

Excluding those black circles, we may roughly state that from 16 to 18 meters or the mean plus one standard deviation, for example in E, 10.79 + 5.71 = 16.5 meters (Table 7) which is expected to include 84 % of cases in a theoretical normal distribution, could be taken as the substantial mating distance in *Altingia excelsa*. If we assume 16 or 18 meters as the "mating distance", the area, as a circle, of a family clump in the species is roughly estimated to be $\Pi r^2 = 200\ m^2$ with $2r = 16$ or 18 meters.

Granting that 16 or 18 meters as the mating distance, we encounter a new problem. We have already concluded that B and C families are genetically isolated from each other in electrophoretic as well as in several leaf characters. In Figure 3, however, it is shown that the two families are growing rather close to each other in the forest. Why are they genetically isolated in spite of their standing in neighborhood? Table 8 shows the intertree distances among six trees, excluding the farthest one of B family, and between those of B and three of C. We find in Table 8 that the intertree distances between the two families are from 5 to 20 meters, but 17 out of 18 or 94 % was less than 17.5 meters, which are understood as to be within the range of the mating distance.

Table 8. Intertree distances of six trees* in family B and distances between three trees of family C and six of family B

	Intertree distance (m)								Total	Mean	SD
	0.1-2.5	2.6-5.0	5.1-7.5	7.6-10.0	10.1-12.5	12.6-15.0	15.1-17.5	17.6-20.0			
Within B	2	3	3	4	1	2	0	0	15	7.09	3.71
Between B and C	0	0	3	3	3	4	4	1	18	12.25	3.98

* One farthest tree was discarded.

Why are trees of B and C families isolated sexually in spite of growing within the range of mating distance? At present, we have no data on the problem, but we guess that the two families may not flower at the same time of a year inducing sexual isolation between them. The argument will be given in the section of DISCUSSION AND CONCLUSION to follow.

II. *Agathis borneensis* Warb.

Actual observations were compared with expected distribution of disagreement counts of isoperoxidases in individual trees (Table 9 and Figure 5).

From Table 9 we see that number of actual observations appeared to exceed expectations of the class of 0 and 1, and also that of 6 or more of the disagreement counts, though the excess being not enough to reach the level of statistical significance. In Figure 5, we find that the general situation of the relation between two curves is similar with that of *Altingia* given in Figure 2. It is thus considered that inbreeding probably occurs to some extent in *Agathis borneensis*, too.

Table 9. Comparison between theoretical expectations and actual observations of disagreement counts in <u>Agathis borneensis</u>

	Disagreement count						Total
	0,1	2	3	4	5	6≦	
Expectation	20.4	39.5	53.7	48.1	29.8	18.3	
Observation	28	34	45	42	33	28	210

$\chi^2 = 11.01$ d.f. = 5 P = 0.05 - 0.10

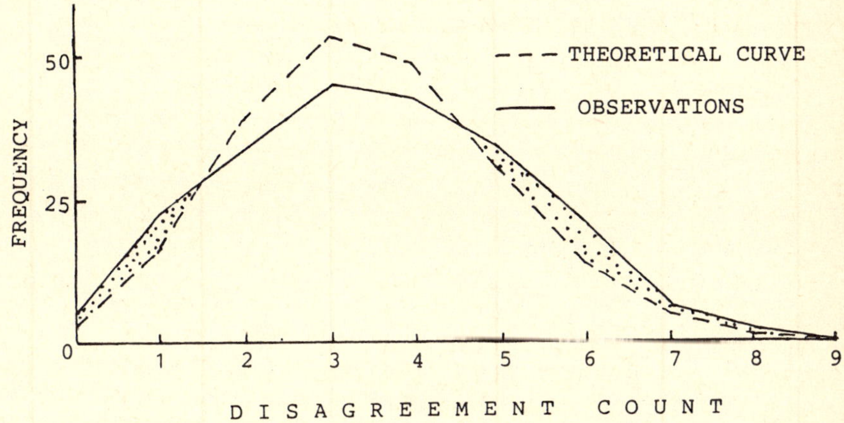

Figure 5. Comparison between observed and expected frequencies of disagreement counts in <u>Agathis borneensis</u>.

Next problem to be solved is if the occurrence of lower disagreement counts is related to small intertree distance. The results of investigation is given in Table 10, from which we find that in <u>Agathis borneensis</u> disagreement counts are randomly distributed quite independent of the intertree distance. Thus, we come to conclude that in <u>Agathis borneensis</u> parental and child trees do not form family clumps in a natural stand. Since the species is monoecious, occurrence of inbreeding in the species may probably be from self-fertilization of individual trees.

DISCUSSION AND CONCLUSION

In this section, argument will be limited to essential points. The full accounts with citations will be published on another occasion as mentioned in the first page of this paper.

Table 10. Comparison between theoretical expectations and actual observations of disagreement counts in *Agathis borneensis*

Intertree distance (m)		Disagreement count 0,1	2,3	4,5	6 ≦	Total
0 - 10	Expectation	4.50	13.00	12.84	4.67	
	Observation	5	16	8	6	35
11 - 20	Expectation	6.56	18.94	18.71	6.80	
	Observation	10	18	16	7	51
21 - 30	Expectation	5.66	16.34	16.13	5.86	
	Observation	3	22	16	3	44
31 ≦	Expectation	10.29	29.72	29.34	10.67	
	Observation	9	22	37	12	80
	Total	27	78	77	28	210

$\chi^2 = 12.34$ d.f. = 9 P = 0.10 - 0.20

It is needless to say that plants growing in a population would naturally involve related and non-related individuals more or less mingled with each other. If, under the control of some biological or environmental conditions, mating occurs mainly among related plants, but very seldom among non-related ones, intra-population differentiation in genetic characters would arise, though the population may be attended with some disadvantages due to inbreeding depression. What would be the facts in the tropical tree species investigated in the present study.

On the understanding that the degree of variation among trees in the pattern of their isoenzymes could most probably manifest dissimilarity in their genotypic constitution, the disagreement counts are used in the present study to measure genotypic relationship among individual trees. The genotypic dissimilarity thus measured would involve the combined effect of dissimilarities in gene frequency and genetic association involving linkage disequilibrium, if any. It will be needless to say that genotypic constitution is more similar among consanguineous trees than among non-consanguineous ones.

In *Altingia excelsa* growing in a natural forest of Ciwidey, West Java, occurrence of inbreeding and growing of consanguineous trees in the neighborhood have been evidenced from the distribution patterns of disagreement counts in Figure 2 and the χ^2 test of Table 3. In *Agathis borneensis*, on the contrary, inbreeding appears to be actually occurring but less frequently than in *Altingia* and no cluster formation by consanguineous trees has been observed.

On the basis of disagreement counts combined with their distribution in the stand, five groups of trees, presumably family clumps, have been identified in Altingia (A to E in Figure 3). Further examination has proved that in these five groups, inter-group differentiation in isoperoxidase components (Table 4) as well as in five size characters of leaves and petioles are apparent (Tables 5 and 6).

Measurement of intertree distances in family clumps in Altingia excelsa has shown that 16 to 18 meters are considered to be the mating distance for the species, and the area one family occupies has been estimated to be approximately 200 to 250 m².

Notwithstanding the genetic differentiation in isoperoxidase components and vegetative characters among five family clumps as described above, we find that some family clumps are spatially not separated far enough from each other. We can mention B and C groups, trees of which are standing rather close to each other within the mating distance in spite of their genetic differentiation. Why have the two groups been genetically isolated?

We recollect in this connection that in general, flowering time of a plant is controlled by environmental conditions on the one hand, and genetic constitution of the plant on the other. Representative of the environmental conditions may be the photoperiodic and vernalizing effect of seasons, and the lack of seasonality in the tropical forest leads to irregularity and lack of coincidence in flowering among not only related species, but also among individual trees of the same species. Thus, in the tropical rain forest where climatic control is absent, genetic make-up of individual trees should be responsible for the initiation of their flowering. Here it may be needless to say that trees of the same family are likely to possess many of the same genes in common. So, it would be quite natural to expect that they will show more or less synchronized flowering among them, but not with trees of the other family. This might be the reason why two families B and C are sexually isolated from each other though their trees are distributed within the mating distance. In such a way, different family clumps in Altingia excelsa would be subjected to the effect of genetic drift. Additional investigation is desirable.

In Agathis borneensis it is indicated that inbreeding occurs to some extent, but there is no sign of clustering of a group of trees with low disagreement counts or formation of family clumps as detected in Altingia. It is inferred that related trees are dispersed widely in the forest without forming a small group per family. The inbreeding effect observed should then most probably be due to self-fertilization of individual trees.

A MULTILOCUS STUDY OF NATURAL POPULATIONS OF PINUS SYLVESTRIS

Outi Muona and A.E. Szmidt

ABSTRACT

Genetic variation of *Pinus sylvestris* was studied in three natural populations in northern Sweden (latitudes 66°40'N, 65°30'N, and 64°30'N). From seed harvests of each population the macrogametophyte and embryo of 133-134 seeds were studied with respect to 14 variable enzyme loci. The degree of genetic differentiation between the populations was very low (G_{ST}=0.006). The average fixation indices in the populations were 0.135, 0.092, and 0.085, respectively. These positive fixation indices are probably mostly due to selfing. Gametic disequilibrium between the loci was studied in zygotes and separately in the ovule and pollen pools. Disequilibrium was also divided into between and within individual components. Small, but significant disequilibria were found in the ovule pool of the northernmost population. The between individual component was also significant in this population. As the significantly associated locus pairs were unlinked, we concluded that the disequilibria are due to restricted effective population size. The decrease in the effective size may be due to uneven seed production, partial selfing and different male and female numbers, which are all known to occur in populations of *Pinus sylvestris*.

INTRODUCTION

The genetic structure of coniferous forest trees has been studied extensively in recent years (see Mitton 1983 for a review). Conifers have been shown to harbor large amounts of genetic variability, but little geographic differentiation has been found in allelic frequencies at enzyme loci. On the other hand, microdifferentiation has been observed in many cases.

Despite much work in this field, there have been hardly any studies on associations between loci in conifer populations, even though other organisms have been extensively studied. The interest in multilocus systems was at first mainly in demonstrating the effects of epistatic natural selection (see Lewontin 1974). Later theoretical work has shown that many factors, including drift, population subdivision, and selection on other loci may give rise to disequilibria between neutral loci (see Hedrick *et al.* 1978 for a review). The general finding has been that associations are rare in random mating organisms, except when recombination is highly restricted, whereas disequilibrium has been common in predominantly selfing plant populations (Hedrick *et al*. 1978, Brown 1979). It has often been very difficult to assign causes to the observed disequilibria. In some respects, conifers are interesting material for such studies. The mating system deviates from random mating because of partial selfing and variation in male and female gamete production. Additionally, conifers have special features which permit detailed study of their population genetic structure. Simultaneous study of endosperms and embryos in the seed allows the separation of female and male gametic contributions to the zygote. It is thus possible to study separately the ovule and pollen pools that make up the zygotes. This feature has been used extensively in mating system and seed orchard studies (see Mitton 1983 and Adams 1983 for reviews). It can also be used in studying multilocus associations in conifer populations by measuring the disequilibria separately in ovules and pollen, and by studying components within and between individuals separately (see Weir 1979).

Here we describe the genetic structure of the zygote populations of three natural populations of Scots pine, *Pinus sylvestris*, in northern Sweden. We first discuss general features: the degree of genetic differentiation of populations and the level of inbreeding. We then compare the ovule and pollen pools that make up the zygotes, and finally proceed to a multilocus analysis of the populations.

MATERIAL AND METHODS

Scots pine seed was collected from three natural populations in northern Sweden by the Institute of Forest Improvement at Sävar in 1970. The populations were Jokkmokk (latitude $66^{\circ}40'$N, longitude $20^{\circ}00'$E, elevation 140 m), Norrsjö ($65^{\circ}30'$N, $16^{\circ}59'$E, 200 m) and Tallsjö ($64^{\circ}30'$N, $17^{\circ}00'$E, 150 m). The purpose of the collection was to provide big seedlots for extensive field testing. The collection was done as for commercial purposes. According to collection guidelines, more than 50 trees are to be included in such collections. We obtained samples from these seedlots and regard them as random samples of the zygote population. Our sample sizes were 134, 134 and 133, respectively. The embryo and the macrogametophyte of seeds were analysed simultaneously by starch gel electrophoresis. Sixteen loci were analysed, of which 14 polymorphic loci are discussed in this study. The enzymes studied were acid phosphatase (APH), aconitase (ACO), fluorescent esterase (FE), glutamate dehydrogenase (GDH), glutamate-oxaloacetate-transaminase (GOT), leucineaminopeptidase (LAP), malate dehydrogenase (MDH), 6-phosphogluconate dehydrogenase (6PGD), phosphoglucose isomerase (PGI), and shikimate dehydrogenase (SDH). The methods and relevant references for inheritance of these allozyme markers are given by Szmidt and Muona (1985).

For analyzing population differentiation, we used Nei's (1973) gene diversity statistics. The measure G_{ST} estimates what proportion of total heterozygosity is due to between population differences. Single locus fixation indices were computed based on total observed and total expected heterozygosities, as described by Curie-Cohen (1982). Between locus heterogeneity was studied with X^2-tests. Allelic frequencies were compared with the G-test (Sokal and Rohlf 1981). Expected heterozygosities and fixation indices were compared between populations with paired t-tests.

For multilocus analyses, diallelic loci were used. When there were more than two alleles at a locus, all but the most frequent allele were pooled to form a synthetic allele. In most cases, the third most frequent allele was rare. This method does not result in much loss of information, but the computations are simpler and testing is easier. Several measures of disequilibrium were computed following Cockerham and Weir (1977) and Weir (1979). Let us consider two loci, A and B, with alleles A,a and B,b. The usual measure of gametic disequilibrium between loci A and B is

$$D = P^{AB}_{..} - p_A p_B$$

where $P^{AB}_{..}$ is the frequency of AB gametes, and p_A and p_B the frequencies of the alleles A and B, respectively. For the estimation of D in non-random mating populations, gametic data are needed. Maximum likelihood estimates of D were obtained in all populations from the whole set of gametes (D), and separately for the ovule (D_{ov}) and pollen pools (D_{pol}), following Weir (1979). We tested whether estimates of D differ significantly from zero as described by Cockerham and Weir (1977, p.144). D_{ov} and D_{pol} were tested with the usual X^2-test (see Weir 1979, p. 246).

D depends on allelic frequencies, and to lessen this dependence, it is useful to standardize D by computing a correlation coefficient

$$r = D/\sqrt{p_A p_a p_B p_b}.$$

Such standardization was made for D, D_{ov} and D_{pol} to yield r, r_{ov} and r_{pol}.

D is a composite measure, which can be partitioned into a between individual and a within individual component

$$D_b = P^{A.}_{.B} - p_A p_B \qquad \text{and} \qquad D_w = P^{AB}_{..} - P^{A.}_{.B}$$

The sum of these two components is D. D_b compares the frequency of the allelic combination AB in two gametes uniting to form a zygote ($p^{A.}_{.B}$) with the frequency of the AB combination in two random alleles ($p_A \cdot p_B$). It is non-zero when there are departures from the random union of gametes, which may be caused by non-random mating, allelic frequency differences between males and females, sampling effects, or selection (see Cockerham and Weir, 1977). The alleles A and B in the one individual may be both from one parent (frequency $P^{AB}_{..}$), or the individual may have received them from different parents (frequency $P^{A.}_{.B}$). The within individual component compares these two frequencies. Note that these two components can only be distinguished when coupling and repulsion heterozygotes can be separated. A non-zero value of D can be due to either component, or both. It is also possible that non-zero components of opposite sign cancel each other to result in a non significant D. The significance of these components was tested with X^2-tests described by Cockerham and Weir (1977).

For each population, estimates of r, r_{ov}, r_{pol}, D_b and D_w for the 91 locus pairs were obtained. The distributions of the estimates were compared between populations using an analysis of variance. The variances in different populations were compared with Levene's robust test for equality of variances (Brown and Forsythe 1974), computed by BMDP 81 statistical package (Dixon 1981).

The linkage relationships of the set of loci are partially known. Some of these loci have been studied by Rudin and Ekberg (1978) and a more extensive study was conducted by Szmidt et al. (1985). Of the 14x13/2=91 locus pairs, 82 are known to be unlinked or at most very loosely linked. The recombination frequency between GOT-B and LAP-B is about 20%, the rest of the known recombination frequencies are more than 33%. The possible linkage of nine pairs is not known.

RESULTS

Genetic differentiation of populations. The allelic frequencies in the different populations are given in Table 1. The sample sizes were quite large (about 268 alleles/locus), and some of the allelic frequency differences were significant, even if rather small. Between the populations of Jokkmokk and Norrsjö, the loci GOT-B, 6PGD-A and 6PGD-B differed, between Jokkmokk and Tallsjö, the loci GOT-B, SDH-B, 6PGD-A, 6PGD-B, and PGI-B. Between the Norrsjö and Tallsjö populations, only LAP-A differed. Thus, the northermost Jokkmokk population differed from the other two to some extent. However, if we consider the G_{ST}-values (Table 2), it is evident that the degree of between population differentiation was low. Even the highest value of G_{ST}, for the locus 6PGD-A, was only 0.017, and the average was 0.006.

The genic diversities in different populations and loci are listed in Table 2. These results are not comparable to other studies, because this is not a random set of loci: only variable loci were chosen for this study. The populations can be compared to each other, however. Despite the small differences in allelic frequencies, the northernmost Jokkmokk population had consistently (at ten of the 14 loci) a higher degree of expected heterozygosity than the other two populations. The difference between the populations of Jokkmokk and Norrsjö was statistically significant in a pairwise t-test (t_d = 2.47, $P < 0.05$), as well as that between Norrsjö and Tallsjö (t_d = 2.90 $P < 0.05$). The two extreme populations differed most (t_d = 4.22 $P < 0.001$).

	Jokkmokk	Norrsjö	Tallsjö
LAP-A	134	134	132
1	.045	.022	.000
2	.899	.948	.981
3	.041	.019	.011
4	.015	.011	.008
LAP-B	134	134	131
1	.015	.015	.004
2	.922	.940	.958
3	.060	.037	.031
4	.004	.007	.008
GOT-B	134	134	133
1	.228	.097	.086
2	.287	.399	.380
3	.485	.504	.534
APH	133	132	131
1	.000	.008	.008
2	.011	.004	.000
3	.857	.860	.901
4	.124	.110	.080
5	.008	.015	.011
6	.000	.004	.000
FE	132	134	133
1	.727	.769	.767
2	.152	.112	.094
3	.102	.119	.128
4	.019	.000	.011
SDH-A	134	134	133
1	.000	.026	.023
2	.851	.813	.846
3	.097	.101	.083
4	.052	.066	.049
SDH-B	130	134	133
1	.058	.030	.015
2	.004	.011	.000
3	.938	.959	.985
GDH	130	132	131
1	.400	.413	.363
2	.600	.587	.637
6PGD-A	126	122	129
1	.563	.713	.659
2	.437	.287	.341
6PGD-B	126	122	129
1	.690	.770	.779
2	.290	.230	.221
3	.020	.000	.000
MDH-A	134	134	133
1	.067	.063	.075
2	.933	.937	.925
MDH-B	132	130	131
1	.659	.673	.710
2	.341	.327	.290
PGI-B	134	134	133
1	.022	.011	.000
2	.000	.011	.008
3	.929	.959	.977
4	.049	.019	.015
ACO	133	134	132
1	.004	.015	.027
2	.944	.925	.939
3	.053	.060	.034

Table 1. Allelic frequencies in three natural populations of Scots pine at 14 enzyme loci.

	Jokkmokk	Norrsjö	Tallsjö	G_{ST}
LAP-A	.187	.101	.037	.015
LAP-B	.147	.114	.081	.004
GOT-B	.630	.577	.563	.012
APH	.250	.248	.182	.003
FE	.437	.382	.387	.003
SDH-A	.264	.324	.275	.002
SDH-B	.116	.079	.030	.009
GDH	.480	.485	.462	.002
6PGD-A	.492	.409	.449	.017
6PGD-B	.439	.354	.344	.007
MDH-A	.125	.119	.139	.001
MDH-B	.449	.440	.412	.002
PGI-B	.134	.080	.044	.008
ACO	.107	.140	.116	.002
Average over				
14 loci	.304	.275	.252	.006

Table 2. Genic diversities and genetic differentiation between three natural populations at 14 enzyme loci.

Inbreeding. The fixation indices of the single loci in different populations are listed in Table 3. All populations had relatively high positive fixation indices on the average (.13, .092, .085), but there was some variability between loci. The heterogeneity was statistically highly significant ($P < 0.001$) in Jokkmokk, where it was due to one locus, PGI-B. When this locus was removed, the rest of the loci were homogeneous. In Norrsjö, the lesser heterogeneity ($P < 0.05$) was due to two loci, PGI-B and SDH-B. When the fixation indices are compared among populations, the different loci give rather consistent results: Jokkmokk had the highest fixation index in eight of the 14 cases. The differences, as measured by a pairwise t-test, were significant between Jokkmokk and Norrsjö (t_d = 2.24, $P < 0.05$). The Tallsjö population had a lower average fixation index than Norrsjö, but still did not differ significantly from Jokkmokk because of more variability between loci (t_d = 2.08 $P < 0.058$). Norrsjö and Tallsjö had similar fixation indices.

	Jokkmokk	Norrsjö	Tallsjö
LAP-A	.244	.259	.391
LAP-B	.136	.085	.154
GOT-B	.266	.147	.092
APH	.187	.115	.077
FE	.082	.083	.027
SDH-A	.152	.148	.152
SDH-B	.071	-.034	-.015
GDH	.135	.109	.092
6PGD-A	.258	.119	.172
6PGD-B	.060	.119	-.058
MDH-A	.047	.058	.027
MDH-B	.090	.074	.073
PGI-B	-.059	-.029	-.018
ACO	.226	.040	.018
Average over			
14 loci	.135	.092	.085

Table 3. Fixation indices in three natural populations at 14 enzyme loci.

Comparison of ovule and pollen gene pools. The separate male and female contributions to the zygote can be traced, because the maternal alleles can be identified in the haploid macrogametophyte. This is important, because the mother and father populations of the zygotes may not be the same, which may in turn cause differences in male and female allelic frequencies. We compared the frequencies of males and females in the three populations, but found very few differences. In Jokkmokk, there was a highly significant difference at the locus FE ($G = 15.4$ P 0.001), in Norrsjö a significant difference at MDH-B ($G = 5.15$ P 0.01), and in Tallsjö no differences at all. We also compared the genic diversities of ovule and pollen pools (listed in Table 4), and the results were, quite naturally, similar to those in zygotes. For both ovules and pollen, the Jokkmokk population had the highest diversity. In Jokkmokk, the pollen pool was slightly more genetically diverse than the ovules, but the difference was not statistically significant. In the other two populations, there was no difference. Nor was there any difference in the degree of differentiation of ovules and pollen (Table 4). Since the overall degree of differentiation at these loci was so low, there was not much potential for detecting possible differences between the pollen and ovule gene pools.

	Ovules	Pollen
	H	H
Jokkmokk	0.293	0.312
Norrsjö	0.270	0.275
Tallsjö	0.255	0.250
G_{ST}	0.008	0.009

Table 4. Average genic diversities and genetic differentiation of ovule and pollen pools in three natural populations of Scots pine.

Gametic disequilibrium. The number of disequilibrium coefficients for the 91 locus pairs in three populations is so large that it is not possible to present estimates of all of them. Table 5 presents a summary of the numbers of observed significant disequilibria of various kinds in the different populations. Some significant values are expected because of the large number of the tests. For the gametic coefficient, D, there were 12 significant values in Jokkmokk, whereas the other two populations had fewer significant values, 3 and 5, respectively. It is usually very difficult to interpret the nature of disequilibria, because of the variety of possible causes (see e.g. Avery and Hill 1979). Epistatic selection alone can not maintain disequilibrium, but some restriction of recombination is required either through linkage or inbreeding. In our data, all of the significantly associated pairs are known to be unlinked or at most loosely linked. In a nearly random mating population, epistatic selection can be excluded as a possible cause of disequilibrium. Additionally, in the different populations, different locus pairs were involved in the significant disequilibria, except for one locus pair. Also, the absolute values of the significant disequilibria were quite small, as seen in Fig.1. These results point to that the significant disequilibria were due to population structure effects, such as reduced effective population size.

These disequilibria can be studied in more detail by considering the components between (D_b) and within (D_w) individuals. Columns 2 and 3 in Table 5 show that the between individual disequilibrium, D_b, was more often significant than D_w in these populations. In Jokkmokk, D_b was significant in nine cases and D_w in six cases. The other two populations did not have many significant D values, and the components did not reveal additional associations.

	D	D_b	D_w	D_{ov}	D_{pol}
Jokkmokk	12	9	6	11	4
Norrsjö	3	2	2	5	6
Tallsjö	5	2	0	6	3

Table 5. Number of significant disequilibria among 91 locus pairs in three natural populations of Scots pine. See text for explanations.

A further possibility of more detailed study is to consider the two sets of gametes separately. These results are also shown in Table 5 (columns 4 and 5). The picture from maternal gametes is similar to that found earlier: there is some evidence of disequilibrium in the Jokkmokk population (11 significant values), much less in the other two populations. For pollen gametes, there was no clear difference between the populations. In the southern populations, there was no difference in the amount of disequilibrium between the ovule and pollen pools; both showed very little evidence of associations. However, in the Jokkmokk population, the ovule and pollen pools clearly differed: there was hardly any disequilibrium in the pollen pool, but much in the ovule pool. In general, different pairs of loci were associated in ovules and pollen.

In all these tests, one has to take into account the power of the tests. Large sample sizes and intermediate allelic frequencies result in higher power (Brown 1975). The sample sizes were similar in all tests, but there was some difference in allelic frequencies. Here we have to compare the frequencies of the leading allele and the synthetic allele in different populations, because diallelic data are being used. The average diallelic expected heterozygosities were 0.29, 0.27 and 0.24, respectively, in Jokkmokk, Norrsjö and Tallsjö. This means that the tests have largest power in the Jokkmokk population, where we, in fact, found most disequilibrium. However, the differences are not due to this alone, as shown by a comparison of the Jokkmokk ovules and pollen. The pollen pool was slightly more diverse than the ovule pool, but had much less disequilibrium.

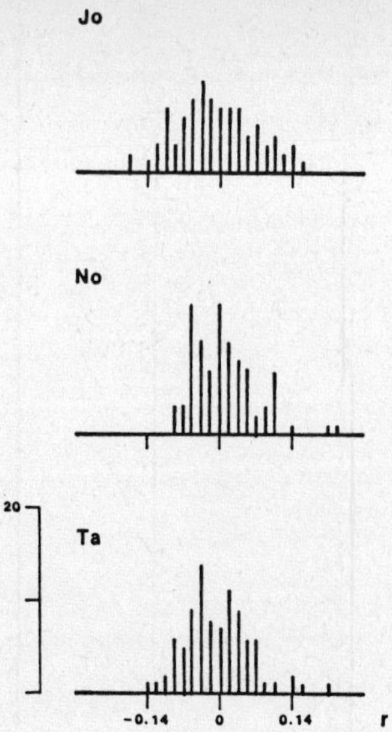

Fig. 1. The distribution of the correlation coefficient between loci (r) in different populations. n = 91. Jo - Jokkmokk, No - Norrsjö, Ta - Tallsjö.

When disequilibria are thought to be due to population structure effects rather than epistatic selection, the distributions, and especially the variances of disequilibria are of more interest than single pairs of loci. We will discuss the distributions of estimates of r, r_{ov} and r_{pol} rather than D, D_{ov} and D_{pol} in order to make comparisons between populations easier after standardization with allelic frequencies. For D_b and D_w, the original estimates are used. The distribution of r is shown in Fig. 1. Note that the distributions can be compared among populations, but not between the coefficients except for r_{ov} and r_{pol}. It should also be pointed out that these distributions are the result of two sampling processes: first the sampling of gametes in the population, and second our sampling of cones and seeds (see Weir and Hill 1980). Since our sample sizes were similar in all populations, it is likely that the differences in the distributions are due to the former sampling. All distributions were centered around zero and in no case did the means differ. The statistic of greater interest here is the variance: reduced effective population size is expected to cause an increase in the variance of disequilibrium. The variances of the distri butions of disequilibrium are listed in Table 6, which also gives the results on comparing the variances in different populations with Levene's test. These results agree with the results based on the number of significant associations. The Jokkmokk population had the largest variance of r. This difference is largely due to the ovule pools, where Jokkmokk again had the highest variance. There was no difference between the variances of the disequilibria in pollen pools. In fact, all gamete pools except for the Jokkmokk ovules had similar variances. Note also that the total gamete pool had a much smaller variance of r than either of the gamete pools, due of course to the twice as large sample size. The results of the components also agree with earlier ones. There was a highly significant difference between the populations in D_b. Jokkmokk had the largest value, the other two populations were approximately similar. The D_w component also differed between populations, even though in no population were there more significant values than expected by chance. However, in Norrsjö and Tallsjö there were much fewer values than expected, 2 and 0, respectively. As these components were not standardized, the differences may be accentuated by allelic frequency differences.

	r	r_{ov}	r_{pol}	D_b	D_w
Jokkmokk	0.0061	0.0128	0.0086	0.000125	0.000202
Norrsjö	0.0040	0.0071	0.0081	0.000066	0.000098
Tallsjö	0.0041	0.0082	0.0074	0.000041	0.000077
$F_{2,270}$	3.40	3.67	0.10	10.38	5.0
$P <$	0.05	0.05	NS	0.001	0.01

Table 6. Variances of correlation coefficients between loci in the total gametic pool, in the ovule pool and the pollen pool, and variances of the between and within individual components of disequilibrium. F- and P-values are from testing the equality of variances.

DISCUSSION

We studied samples of seed obtained from commercial collections. We do not know to what extent single trees have contributed to the collections, but we make the assumption that the differences in these contributions reflect differences in the seed production, which will affect the natural zygote population in a similar way. None of our results point to very limited sampling in any of the locations. However, we still have to be cautious in drawing conclusions.

Genetic differentiation. We observed very little genetic differentiation between these three populations in allelic frequencies. The average degree of differentiation, as measured by G_{ST}, was even lower than has been observed in many other conifers (see Adams 1983). This result agrees with an earlier study by Rudin *et al*. (1974), who reported allelic frequencies from the very north and south of Sweden. At the three loci they studied, we computed the average G_{ST} to be only 0.015. The results on allozymes are in strong contrast to much clinal variation in many adaptive characters of *Pinus sylvestris*. E.g. Mikola (1982) studied bud set phenology in Finnish populations and found significant differentiation within the same latitudinal range that we studied here.

Heterozygosity. Our sample of loci was selected based on variability, and the results are thus not directly comparable to other studies. However, it is evident that all populations contain much variability, with average expected heterozygosities at these loci between 0.25 and 0.30. Even though there was a very low level of differentiation in allelic frequencies, we found statistically significant differences in expected heterozygosity. The northern population had the highest expected heterozygosity, the southernmost the lowest. We do not have an explanation for such differences, our sample of three populations is too small for such considerations.

Inbreeding. We measured inbreeding in these stands by computing fixation indices based on observed and expected total heterozygosity. Such an index is influenced by other factors besides inbreeding, e.g. differences between male and female allelic frequencies, heterogeneous pollen pools, and possibly selection. If male and female allelic frequencies differ, random union of gametes results in heterozygote excess over Hardy-Weinberg expectation. This can be an important factor in conifer populations (Müller-Starck *et al*. 1983). We could at least largely exclude the influence of allelic frequency differences between pollen and ovules at the loci we studied. Among all the 42 tests (3 populations, 14 loci) for differences between males and females, only two were significant. Thus possible differential male and female reproduction is not visible at these loci. Our estimated fixation indices had some between locus variability. This is a common finding (e.g. Shaw and Allard 1982, Hiebert and Hamrick 1983), and may point to that other factors besides inbreeding influence the genotypic structure at the loci. Unfortunately, we do not have estimates of selfing for these populations. When different populations were compared, we found rather consistently that the northernmost population had the highest F. At equilibrium, the relation between F and the selfing proportion is $F_e = (1-t)/(1+t)$, where t (=1-s) is the rate of outcrossing (see Allard *et al*. 1968). The observed fixation indices of 0.13 and 0.09 would then correspond to t = 0.76 and t = 0.84. The estimates of outcrossing in natural stands of *Pinus sylvestris* (e.g. Müller 1977) have usually been higher. Other kinds of inbreeding besides selfing may also contribute to these relatively high fixation indices.

Note that the inbreeding is in embryos. It has been found that in later life stages there is no evidence of inbreeding (no excess homozygosity). Tigerstedt et al. (1982) found that excess homozygosity disappeared rather late, whereas in a study by Yazdani *et al*. (1984) there was no excess homozygosity in 10-20 year old trees. Shaw and Allard (1982) demonstrated this difference between embryos and adults in *Pseudotsuga menziesii*.

Gametic disequilibrium. There are only few other studies on gametic disequilibrium in conifers. Mitton *et al*. (1980) mentioned finding some significant associations in *Pinus ponderosa*, but only five loci were studied. Similarly, Rasmuson (1979) found some cases of association in a study of *Picea abies*, but hardly any in *Pinus sylvestris* among the few locus pairs studied. A special feature of our results is that significant, even if small disequilibria were found between unlinked loci. The general result among predominantly outcrossing and random mating organisms has been that disequilibria are very rare, except when linkage is tight (Brown 1979). In our study, the tests may have been rather powerful, and even small disequilibria were significant. Significant disequilibria between unlinked loci have also been found in laboratory cage populations of Drosophila melanogaster (Langley et al. 1978, Laurie-Ahlberg and Weir 1979). In both cases, such disequilibria could be attributed to finite population size effects in population cages. In random mating populations, low numbers of individuals, and in our study, also variation in reproduction and mating pattern effects, are likely causes of reduced effective population size and thus disequilibria. Selection is an unlikely cause because it can not maintain disequilibrium in the face of so much recombination. Hitchhiking effects are not likely either with unlinked or very loosely linked loci and only little selfing (Hedrick 1980).

As we assume that the disequilibria are due to effective population size influence, it is more important to consider the distribution of disequilibrium rather than to discuss single locus pairs. The distribution of disequilibrium or the correlation coefficient between neutral loci has been studied by many authors (e.g. Hill and Robertson 1968, Hill 1977, Golding and Strobeck 1983). The expected value of disequilibrium is zero, but the variance of the distribution depends on the recombination fraction and the effective population size: the smaller $4N_ec$, the larger the variance (where N_e is the effective population size, c the recombination frequency). Golding and Strobeck (1979) demonstrated that the effect of selfing is to reduce both the effective population size and the effective recombination frequency. Golding (1984) has studied the sampling distribution of linkage disequilibrium, and found that commonly used X^2-tests are adequate for loosely linked loci, but may be inaccurate for tightly linked loci in small populations.

The distributions of disequilibrium are described in Tables 5 and 6 and Fig. 1. We can summarize the results as follows. There was significant disequilibrium in the zygotes of Jokkmokk, where especially the between individuals component was important. When the gametes are considered, we found disequilibria in the total gametic pool and the ovule pool in Jokkmokk, but not in the pollen pool. The other two populations had hardly any evidence of disequilibrium.

The disequilibrium in zygotes due to nonrandom union of gametes can be compared to our results on fixation indices. The Jokkmokk population had the highest fixation index and it also had significant between individuals disequilibrium. The effects are not very large, but inbreeding due to selfing may contribute to the disequilibrium. This component of disequilibrium has been studied only rarely, because it can not be measured in most cases. Laurie-Ahlberg and Weir (1979) estimated D_b in cage populations of D. melanogaster in such cases where there were no double heterozygotes, but found no significant cases, even though there was otherwise much disequilibrium in the populations. Drosophila flies in cages probably mate at random, whereas in pine populations there are some deviations from random mating.

When discussing the disequilibria in the different gametic pools, population size effects should be considered. The variance of disequilibrium between neutral loci depends on population size, and this information can be used for estimating effective population sizes (Hill 1981). This method was used e.g. by Laurie-Ahlberg and Weir (1979). When the sample sizes are small compared to effective sizes, the distribution contains little information on N_e (see Weir and Hill 1980). Following those methods, we obtained very rough estimates of the effective sizes from various gametic pools. We included all locus pairs in the estimates, but used the formulae for unlinked loci, which results in some underestimation. In cases where the variance of disequilibrium is no larger than expected based on our final sampling, no finite estimate can be obtained. The estimated sizes from the total gametic pools were 225 and 1750 for Jokkmokk and Tallsjö, respectively, whereas no estimate could be obtained for Norrsjö. From the maternal gametes, the estimates were 100 and 860 for Jokkmokk and Tallsjö, respectively. From the pollen pools, the estimates for Jokkmokk and Norrsjö were 605 and 1800 with no estimate for Tallsjö. These numbers are quite unexact, but may give some indication of the relative sizes.

Populations may differ from an ideal population in many ways that reduce the effective size from the census number: different numbers of males and females, variation in offspring number, inbreeding, variation of population size over time, and age structure (e.g. Hedrick 1983). Many of these effects have been demonstrated in populations of Pinus sylvestris. Differences in male flowering intensity, differences beween trees in seed production and in the proportion of germinable seed have been documented (Sarvas 1962, Hagner 1965, Koski and Tallkvist 1978, Ryynänen 1982). Nilsson (pers. comm.) has observed that there is much variation in flowering intensity between males and especially between females in the Jokkmokk area. Such effects may be the cause of the somewhat lower effective size and small disequilibria that we observed. However, with our present data set it is not possible to draw detailed conclusions about the differences between populations, because we do not have the details about collection. Differential cone production is an important factor in considerations of effective population size, but its effects may be further accentuated by collection concentrating on trees producing most cones. Our above population size estimates, and the other data (on allelic frequencies, gene diversities etc.), and the information on collection procedures provided by Mr. Ola Rosvall, do not suggest that only a small number of trees had been collected. However, a more specific collection scheme is needed before firm conclusions can be drawn. Still, we suggest that it is possible that occasionally flowering conditions result in a bottleneck, and disequilibria, in a zygotic cohort. In principle, such disequilibria decline rather slowly (see Avery and Hill 1979). However, if small disequilibria are generated only through occasional bottlenecks, they will probably not have much effect on the overall genetic structure of the population.

Acknowledgements. This study has been financed by a grant from the Cellulose Industries Council for Technology and Forest Research to A.E.S. O.M. has received financial support from the Academy of Finland and the Ehrnrooth Foundation. We acknowledge generous help from the staff of the Institute for Forest Improvement at Sävar in providing material.

We thank the participants of the meeting and Dag Lindgren for helpful comments on the study.

REFERENCES

Adams, W.T. 1983. Application of isozymes in tree breeding. In: Tanksley, S.D. and Orton, T.J. (eds.). Isozymes in plant genetics and breeding. Elsevier Science Publishers, Amsterdam, pp. 381-400.

Allard, R.W., Jain, S.K. and Workman, P.L. 1968. The genetics of inbreeding populations. Advances in Genetics 14: 55-131.

Avery, P.J. and Hill, W.G. 1979. Distribution of linkage disequilibrium with selection and finite population size. Genet. Res. Camb. 33:29-48.

Brown, A.H.D. 1975. Sample sizes required to detect linkage disequilibrium between two and three loci. Theor. Pop. Biol. 8:184-201.

Brown, A.H.D. 1979. Enzyme polymorphisms in plant populations. Theor. Pop. Biol. 15:1-42.

Brown, M.G. and Forsythe, A.B. 1984. Robust tests for the equality of variances. J. Amer. Stat. Assoc. 69:364-367.

Curie-Cohen, M. 1982. Estimates of inbreeding in natural population: a comparison of sampling properties. Genetics 100:339-358.

Cockerham, C.C. and Weir, B.S. 1977. Digenic descent measures for finite populations. Genet. Res. Camb. 30:121-147.

Dixon, W.J. (ed.) 1981. BMDP Statistical software 1981. Univ. Calif. Press, Los Angeles.

Colding, G.B. 1984. The sampling distribution of linkage disequilibrium. Genetics 108: 257-274.

Golding, G.B. and Strobeck, C. 1979. Linkage disequilibrium in a finite population that is partially selfing. Genetics 94:777-789.

Golding, G.B. and Strobeck, C. 1983. Two-locus, fourth-order gene frequency moments: implications for the variance of squared linkage disequilibrium and the variance of homozygosity. Theor. Pop. Biol. 24:173-191.

Hagner, S. 1965. Cone crop fluctuations in Scots pine and Norway spruce. Studia Forest. Suecica 33:1-21.

Hedrick, P.W. 1980. Hitchhiking: a comparison of linkage and partial selfing. Genetics 94:791-808.

Hedrick, P.W. 1983. Genetics of Populations, Science Books International, Boston.

Hedrick, P.W., Jain, S.K. and Holden, L.R. 1978. Multilocus systems in evolution. Evol. Biol. 11:101-184.

Hiebert, R.D. and Hamrick, J.L. 1983. Patterns and levels of genetic variation in Great Basin bristlecone pine, Pinus longaeva. Evolution 37:302-310.

Hill, W.G. 1977. Correlation of gene frequencies between neutral linked genes in finite populations. Theor. Pop. Biol. 11:239-248.

Hill, W.G. 1981. Estimation of effective population size from data on linkage disequilibrium. Genet. Res. Camb. 38:209-216.

Hill, W.G. and Robertson, A. 1968. Linkage disequilibrium in finite populations. Theor. Appl. Genet. 38:226-231.

Koski, V. and Tallkvist, R. 1978. Tuloksia monivuotisista kukinnan ja siemensadon määrän mittauksista metsäpuilla, Summary: Results on long-time measurements of the quantity of flowering of forest trees, Folia For. 364:1-60.

Langley, C.H., Smith, D.B. and Johnson, F.M. 1978. Analysis of linkage disequilibria between allozyme loci in natural populations of Drosophila melanogaster. Genet. Res. Camb. 32:215-229.

Laurie-Ahlberg, C.C. and Weir, B.S. 1979. Allozymic variation and linkage disequilibrium in some laboratory populations of Drosophila melanogaster. Genetics 92:1295-1314.

Lewontin, R.C. 1974. The genetic basis of the evolutionary process. Columbia Univ. Press, New York.

Mikola, J. 1982. Bud-set phenology as an indicator of climatic adaptation of Scots pine in Finland. Silva Fennica 16:178-184.

Mitton, J.B. 1983. Conifers. In: S.D. Tanksley and T.J. Orton (eds.) Isozymes in plant genetics and breeding, Part B. Elsevier Science Publishers, Amsterdam, pp. 443-472.
Mitton, J.B., Sturgeon, K.B. and Davis, M.L. 1980. Genetic differentiation in ponderosa pine along a steep elevational transect. Silvae Genetica 29:100-103.
Müller, G. 1977. Untersuchungen über die natürliche Selbsbefruchtung in Beständen der Fichte (Picea abies (L) Karst.) und Kiefer (Pinus sylvestris). Silvae Genetica 26: 207-217.
Müller-Starck, G., Ziehe, M. and Hattemer, H. 1983. Reproductive systems in conifer seed orchards. 2. Reproductive selection monitored at an LAP gene locus in Pinus sylvestris L. Theor. Appl. Genet. 65:309-316.
Nei, M. 1973. Analysis of gene diversity in subdivided populations. Proc. Natl. Acad. Sci. USA 70:3321-3323.
Pulliainen, E. and Lajunen, L. 1984. Chemical composition of Picea abies and Pinus sylvestris seeds under subarctic conditions. Can. J. For. Res. 13:214-217.
Rasmuson, M. 1979. Interactions between loci - models and observations. In: Rudin, D. (ed.) Proc. Conf. Biochem. Genet. Forest Trees, Gotab, Stockholm. pp. 62-71.
Rudin, D. and Ekberg, I. 1978. Linkage studies in Pinus sylvestris L. -using macro gametophyte allozymes. Silvae Genetica 27:1-12.
Rudin, D., Eriksson, G., Ekberg, I. and Rasmuson, M. 1974. Studies of allele frequencies and inbreeding in Scots pine populations by the aid of the isozyme technique. Silvae Genetica 23:10-13.
Ryynänen, M. 1982. Individual variation in seed maturation in marginal populations of Scots pine. Silva Fennica 16:185-187.
Sarvas, R. 1962. Investigations on the flowering and seed crop of Pinus sylvestris. Comm. Inst. For. Fenn. 53:1-198.
Shaw, D.V. and Allard, R.W. 1982. Isozyme heterozygosity in adult and open-pollinated embryo samples of Douglas-fir. Silva Fennica 16:115-121.
Sokal, R.R. and Rohlf, F.J. 1981. Biometry. 2nd ed. Freeman, San Francisco.
Szmidt, A.E. and Muona, O. 1985. Genetic effects of Scots pine domestication. Submitted, this volume.
Szmidt, A.E., Muona, O. and Yazdani, R. 1985. Linkage relationships in Scots pine (Pinus sylvestris L.). (in prep.)
Tigerstedt, P.M.A., Rudin, D., Niemelä, T. and Tammisola, J. 1982. Competition and neighbouring effect in a naturally regenerating population of Scots pine. Silva Fennica 16:122-129.
Weir, B.S. 1979. Inferences about linkage disequilibrium. Biometrics 35: 235-254.
Weir, B.S. and Hill, W.G. 1980. Effect of mating structure on variation in linkage disequilibrium. Genetics 95:447-488.
Yazdani, R., Muona, O., Rudin, D. and Szmidt, A.E. 1985. Genetic strucure of a Pinus sylvestris L. seed tree stand and naturally regenerated understory. Forest Science (in press).

GENETIC EFFECTS OF SCOTS PINE (PINUS SYLVESTRIS L.) DOMESTICATION

Alfred E. Szmidt and Outi Muona

ABSTRACT

Genetic variation at 13 polymorphic enzyme loci was studied in progenies from seed orchards and natural populations of Pinus sylvestris in Sweden. Progenies were analysed at two life stages: embryos and 9-11 year old trees. Measures of genetic diversity for the 13 loci revealed much variability in the progenies studied. There was no visible reduction of the genetic diversity in the orchard progenies compared to progenies from natural populations. The average fixation indices in embryos were high and positive, probably due to the occurrence of inbred individuals. A trend of higher fixation indices in the progenies from the northern seed orchards and natural populations was observed. Increased selfing due to scarce and uneven flowering in the northern populations could account for this result. In addition, high fixation indices found in the northernmost seed orchard could also result from the restricted pollen flow between the blocks in this orchard. The young trees planted 9-11 years ago in the field had low fixation indices indicating that excess homozygosity observed in embryos had been eliminated during later stages of development.

INTRODUCTION

Man-managed plant populations are subject to considerable alteration of their original genetic makeup (Harlan 1975). As compared to agricultural crop plants, forest tree populations are still at a very early stage of the domestication process (Libby 1973). However, the rapid development of breeding techniques in forestry is expected to accelerate this process. The impact of domestication processes upon the genetic structure of cultivated forests will depend on the breeding techniques used. The most rapid alterations will occur following a large scale vegetative propagation of selected individuals. Other breeding techniques based on sound strategies are not expected to bring about such abrupt genetic changes. In Sweden, about 40 % of the plant material used for Scots pine reforestation represents open-pollinated seed orchard progenies (Skogsstatistisk Årsbok 1982). As the newly established orchards reach seed production age, this value is expected to increase substantially. A similar trend exists for many other commercially important forest tree species (Adams 1981, Moran et al. 1980). The above situation has resulted in a growing demand for studies aimed at monitoring of the possible genetic changes in cultivated forest tree populations (Adams 1981). In the present study, we monitored genetic variation in progenies from natural populations and seed orchards of Scots pine by means of isozyme analyses. The aim of this study was to determine whether the use of seed orchard progenies affects levels of gene diversity for future plantations of this species. In addition, we also attempted to measure changes in the genetic structure of these progenies between two different life phases, i.e. embryos and 9-11 years old trees.

MATERIAL AND METHODS

Between 1971 - 1975 the Institute of Forest Improvement at Sävar (Sweden) established an extensive series of field experiments with open-pollinated progenies from 10 seed orchards and 7 natural populations in Scots pine. Each seed orchard included in this series was represented by progenies from three consecutive seed crops: 1971, 1972 and 1973. Nursery-raised seedlings were planted in the 15 field trials distributed in northern Sweden (Ericsson and Hadders 1975, Ericsson 1980). The aim of this experiment was to compare the field performance of seed orchard progenies with those from natural populations in order to determine and verify boundaries for seed transfer in Sweden. In our study, progenies from three seed orchards, each represented by three consecutive seed crops, as well as progenies from three natural populations were analysed. The locations of origin of the material are shown in Fig. 1. Additional data on the populations are given in Tables 1 and 2. At the time of seed collection, all three seed orchards were mature and

Seed orchard	Area (ha)	Initial clone number	Average clone age in 1971 (years)	Average latitude of origin	Average elevation of origin (m)
Hortlax (Ho)	8.1	67	11 - 16	66°17'	312
Domsjöänget (Do)	6.0	52	16	64°30'	334
Alnön (Al)	14.5	40	12	64°30'	342

Table 1. Some characteristics of the Scots pine seed orchards analysed in this study.

Population	Latitude	Longitude	Elevation (m)
Jokkmokk (Jo)	66°40'	20°00'	140
Norrsjö (No)	65°30'	16°59'	200
Tallsjö (Ta)	64°30'	17°00'	151

Table 2. Locations and elevations of the natural populations of Scots pine.

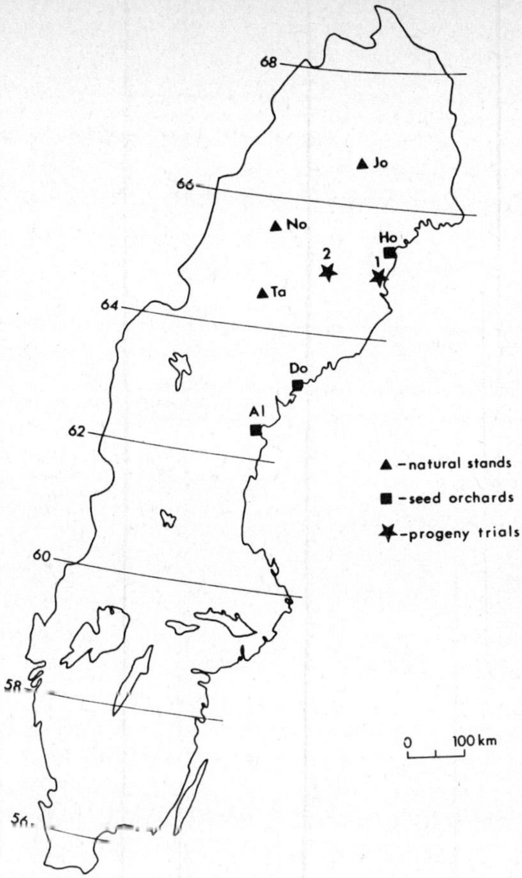

Figure 1. Location of the seed orchards, natural populations, and progeny trials.

in full seed production. Approximately 40 % of the clones in Domsjöänget and Alnön were identical. The clones in Hortlax were entirely different and of more northern origin. In Hortlax, each orchard block (ca 0.8 ha) was surrounded with a 7 m high hedge consisting of birch and Norway spruce. This was not the case in the other two orchards. Two life stages were studied: germinating embryos isolated from seeds and 9-11 year old plants growing in the field trials Drängsmark (1) and Raggsjö (2) (see Fig. 1). In both trials, progenies from each seed orchard and each natural population were represented. Seed samples that remained after the establishment of the progeny trials were supplied by the Staff of the Institute of Forest Improvement at Sävar. Seed from natural populations was typical of commercial seed lots, collected from several neighbouring stands. The number of parental trees in each population is unknown. However, seed from no fewer than 50 trees was included in each population sample (Rosvall, personal communication). The seed from seed orchards also represented commercial seed crops collected in three different years. Between 131 to 144 embryos were analysed in each population. To assess the genetic variation in the young trees, winter buds were collected in Raggsjö and Drängsmark.

We shall refer to these samples as Youngs-1 and Youngs-2. From 30 to 44 single trees were sampled in each population in December 1982. The total material represented 36 experimental populations, i.e. nine embryo populations from three seed orchards (with three consecutive seed crops from each), and three embryo populations from natural seed sources. The same set of materials was studied in the young trees from the two field trials. Genetic variation was analysed by means of electrophoretic separation of allozyme markers in starch gels. A total of ten enzyme systems controlled by 16 loci were assayed in embryos. Fifteen of these loci could be analysed in buds from young trees. Due to poor resolution of acid phosphatase isozymes, this locus could not be analysed in bud tissues. Table 3 gives the enzyme systems assayed, numbers of loci scored, and the references concerning the formal genetics.

Enzyme	E.C. Number	Number of loci scored	Reference
Acid phosphatase	3.1.3.2	1	Mejnartowicz (1979)
Aconitase	4.2.1.3	1	Szmidt & Yazdani (in prep.)
Fluorescent esterase	3.1.1.1	1	Yazdani & Rudin (1982)
Glutamate dehydrogenase	1.4.1.2	1	Rudin (unpublished results)
Glutamate-oxaloacetate-transaminase	2.6.1.1	2	Rudin & Ekberg (1978)
Leucineaminopeptidase	3.4.11.1	2	Rudin (1977)
Malate dehydrogenase	1.1.1.37	2	Rudin & Ekberg (1978)
6-Phosphoglucose dehydrogenase	1.1.1.44	2	Szmidt & Yazdani (1983)
Phosphoglucose isomerase	5.3.1.9	2	Szmidt & Yazdani (in prep.)
Shikimate dehydrogenase	1.1.1.25	2	Szmidt & Yazdani (1983)

Table 3. Enzymes assayed, Enzyme Commission numbers, numbers of loci scored and the references for their genetic interpretation.

Allelic frequencies between populations were compared with G-tests (Sokal and Rohlf 1981). Allelic diversity was measured by expected panmictic heterozygosity ($H = 1 - \Sigma p_i^2$) and the average number of alleles ($A = 1/\Sigma p_i^2$). These statistics take into consideration only the allelic diversity. Genotypic distributions were compared to those expected under Hardy-Weinberg conditions with G-tests. They were also characterised with fixation indices, which compare expected and observed heterozygosities. Our estimate of fixation index was based on total observed and total expected heterozygosity, $F = 1 - H_{obs}/H_{exp}$ (see e.g. Curie-Cohen 1982). We did not consider multilocus statistics in this study. Our other studies have shown that there is only little gametic disequilibrium in these natural Scots pine populations (Muona and Szmidt 1985), as well as in the progeny populations (unpublished results). The loci we have considered are in most cases statistically independent. Thus, the single locus measures address adequately the questions on genetic diversity and genotypic distribution.

RESULTS

The data consisted of gene and genotypic frequencies at 15-16 allozyme loci in 36 populations. Because of the extent of the results only summaries of estimates of genetic variation in the populations studied will be presented. Thirteen loci were polymorphic in all embryos and young trees and only these were used for computation of the average genetic parameters discussed. The average number of alleles per locus (A) and the expected panmictic heterozygosity (H) for each population analysed in our study are given in Tables 4 and 5. The average number of alleles per locus appeared to be relatively uniform in the investigated populations and ranged from 2.38 to 2.92 in embryos and 2.31 to 2.63 in the young trees. The lower number of alleles among young trees was probably due to smaller samples from those populations. Comparison of expected heterozygosities did not indicate substantial reduction of diversity in the seed orchard progenies as compared to natural stand progenies. Similarly, no great fluctuation of diversity was found among the seed crops of different years. Average estimates of the observed heterozygosity and fixation indices in the two life stages are presented in Tables 6 and 7. Some characteristic trends were found. The observed heterozygosity was consistently higher in the youngs than in the embryos and ranged from .284 to .364 and .214 to .262, respectively. Seed orchard progenies

Population	Life stage	1971			1972			1973		
		N	A	H	N	A	H	N	A	H
Hortlax	Embryos	144	2.69	.296	135	2.62	.284	132	2.46	.272
	Youngs-1	37	2.38	.316	36	2.46	.293	31	2.31	.304
	Youngs-2	32	2.46	.306	44	2.31	.292	40	2.38	.310
Domsjöänget	Embryos	143	2.92	.305	131	2.85	.264	133	2.62	.260
	Youngs-1	35	2.69	.337	35	2.46	.290	30	2.38	.292
	Youngs-2	34	2.46	.330	42	2.38	.300	38	2.46	.295
Alnön	Embryos	142	2.69	.275	135	2.69	.258	132	2.38	.269
	Youngs-1	31	2.38	.295	34	2.46	.314	30	2.38	.331
	Youngs-2	30	2.46	.291	37	2.54	.299	44	2.38	.300

Table 4. Sample sizes (N), average number of alleles per locus (A), and average expected heterozygosity (H) in the embryos and youngs from annual seed crops from the seed orchards. The values of A and H are based on 13 polymorphic loci.

Population	Life stage	N	A	H
Jokkmokk	Embryos	134	2.92	.308
	Youngs-1	40	2.62	.332
	Youngs-2	30	2.54	.332
Norrsjö	Embryos	134	2.92	.277
	Youngs-1	32	2.85	.316
	Youngs-2	35	2.77	.303
Tallsjö	Embryos	133	2.69	.257
	Youngs-1	31	2.54	.291
	Youngs-2	34	2.54	.279

Table 5. Sample sizes (N), average number of alleles per locus (A), and average expected heterozygosity (H) in the embryos and youngs from the natural populations investigated. The values of A and H are based on 13 polymorphic loci.

Population	Life stage	1971		1972		1973	
		H_o	F	H_o	F	H_o	F
Hortlax	Embryos	.223	.252**	.227	.220*	.216	.234*
	Youngs-1	.311	-.006	.305	-.043	.303	-.017
	Youngs-2	.286	.035	.284	.005	.299	.005
Domsjöänget	Embryos	.244	.144	.214	.185*	.219	.147
	Youngs-1	.332	.006	.295	-.029	.303	-.053
	Youngs-2	.319	.004	.285	.047	.316	-.072
Alnön	Embryos	.245	.119	.239	.061	.238	.117
	Youngs-1	.315	-.069	.337	-.070	.351	-.072
	Youngs-2	.310	-.050	.316	-.057	.322	-.073

Table 6. Averages of observed heterozygosity (H_o) and fixation indices (F) based on 13 polymorphic loci in the embryos and youngs from seed orchards. Results of G-tests for the departures of the observed genotype frequencies from those expected under panmixia:
* - $P < .05$; ** - $P < .01$.

were slightly and consistently less heterozygous than progenies from natural populations. Average fixation indices in all embryo populations were positive indicating in some cases significant homozygote excess over the expected Hardy-Weinberg proportions (Table 6). However, there was also much variation in the fixation indices among populations. The greatest homozygosity excess (F = .220 - .252) was observed in seed crops from the northernmost orchard Hortlax. In the northernmost population Jokkmokk, located at a similar latitude as Hortlax, the fixation indices in embryos were lower. Another characteristic feature was a trend of decreasing fixation indices from the north to the south. The fixation indices found in the embryos from the southernmost orchard Alnön were relatively low, and were similar to those in the embryos from the corresponding natural population Tallsjö. On the other hand, there was little variation in fixation indices among embryos from different annual crops. A comparison of the fixation indices in embryos with those in the corresponding young populations revealed that substantial genetic

Population	Life stage	H_o	F
Jokkmokk	Embryos	.262	.131
	Youngs-1	.364	-.092
	Youngs-2	.336	-.055
Norrsjö	Embryos	.248	.091
	Youngs-1	.337	-.069
	Youngs-2	.317	-.053
Tallsjö	Embryos	.237	.085
	Youngs-1	.328	-.104
	Youngs-2	.294	-.052

Table 7. Averages of observed heterozygosity (H_o) and fixation indices (F), based on 13 polymorphic loci in the embryos and youngs from natural populations.

change had occured between the two life stages. In contrast to embryos, youngs exhibited low, and frequently negative fixation indices, an indication of slight heterozygote excess over panmictic expectation. Thus, it appears that the homozygote excess present at the embryo stage had been removed after few years of field growth.

DISCUSSION

Genetic diversity. Maintenance of high levels of genetic variation in breeding populations of forest trees is an important premise for future selection and improvement. Furthermore, genetic diversity in the cultivated forests appears to be essential for their ability to counterbalance the effects of environmental changes (Kleinschmit 1979). Our results indicate that in spite of the limited number of parents available for crossing in the seed orchards, their progenies still contain an appreciable amount of genetic diversity. Furthermore, no important fluctuations in genetic diversity occurred among different annual seed crops. These results agree well with some other allozyme studies aimed at monitoring gene diversities in seed orchards and wild populations (Adams 1981). Various studies of pollination patterns in forest tree seed orchards indicate that often only a limited number of clones contribute actively to the next generation (Jonsson et al. 1976, Müller-Starck 1982a, 1982b, O'Reilly et al. 1982). Assuming that a similar situation exists in the seed orchards we analysed, it is possible that the amount of genic diversity retained among orchard clones is still greater, though not fully contributed to the offspring. Hence, it appears that by increasing the genetic efficiency of the existing seed orchards the genetic diversity of the offspring may be further enchanced. It should be pointed out, however, that relatively high levels of genetic diversity found in the seed orchard progenies studied here do not entirely exclude the possibility of a loss of variability. As shown recently, rates of contamination with non-orchard pollen can be high in some seed orchards (Smith and Adams 1983, Nagasaka and Szmidt 1985). Dispersal of foreign pollen from surrounding stands into the orchard would tend to increase genetic variability in offspring, and obscure the actual loss of gene diversity. However, these conslusions apply to variation of marker genes. The fate of adaptive quantitative genetic variation may differ because of different selective forces.

Genetic structure of embryos and young trees. The genotypic distributions in the embryo and adult populations of Pseudotsuga menziesii has been studied by Shaw and Allard (1982). A significant excess of homozygotes over panmictic proportions was found in embryos. In contrast, adult populations were close to Hardy-Weinberg equilibrium. Yazdani et al. (1985) compared the genetic structure of three different life stages in a naturally regenerating stand of Pinus sylvestris. They observed that the embryo population contained an excess of homozygotes while young naturally regenerated trees and the adult parental population were in Hardy-Weinberg proportions. Our present results are in good accord with these earlier findings. High and positive fixation indices were found in all embryo populations analysed in this study, indicating a significant excess of homozygous genotypes over panmictic expectation. The most plausible explanation of the excess homozygosity observed in embryos is the occurrence of inbred and selfed individuals. In conifers, selfed individuals are often aborted at very early stages after zygote formation, which frequently results in an increased proportion of empty seeds (Koski 1973). Nevertheless, it is well known that viable selfs can be produced, although they commonly express inbreeding depression and are eventually removed from a population during later life stages (Eriksson et al. 1973, Phillips and Brown 1977). Besides inbreeding, there are also several other factors, which could affect our estimates of the fixation indices, e.g. selection or allelic frequency differences between the ovule and pollen gametic pools. However, according to our estimates the differences between ovules and pollen in these seed orchards and natural stands were quite small and thus should not significantly affect the fixation indices (Szmidt and Muona, unpublished results). Significantly high fixation indices were consistently found in the seed crops from the northernmost orchard Hortlax. In natural populations, the highest fixation indices were from the northernmost population Jokkmokk. On the contrary, the southernmost orchard Alnön and population Tallsjö had lower and more similar fixation indices in embryos. Northern populations of Scots pine are known to flower less abundantly and to produce greater proportions of empty seeds (Koski 1973). The scarcity of pollen supply and the lower number of flowering trees in those populations may lead to increased rate of selfing and, thus, to greater homozygosity. This could explain the high fixation indices in embryos from the northernmost orchard Hortlax and the Jokkmokk population. However, there was also more homozygosity excess in the Hortlax and Domsjöänget orchards than in the natural populations. There is some evidence that conditions of full genetic efficiency such as random mating and equal clone contribution are not met in forest tree seed orchards (Friedman and Adams 1982, Müller-Starck 1982a, 1982b). Most of the earlier allozyme studies have been confined to analysis of seed crops from only a single orchard. However, seed orchards of a particular species are known to differ greatly with regard to a number of features such as age, number and origin of clones, management, location, and the degree of isolation from other pollen sources. Many of these features are likely to have some effect on the genetic composition of the seed crops produced. Thus, estimates obtained from one seed orchard need not apply to many others. The observed variation of fixation indices in the orchard embryos analysed in our study seems to support this suggestion. The Hortlax embryos had higher fixation indices than any other population studied. On the other hand, the fixation indices found in embryos from the southernmost orchard Alnön were of about the same magnitude as in natural populations. Clone age and number are often mentioned as possible factors which may influence the mating patterns in seed orchards. Weak and uneven pollen production in young seed orchards is often reported and this may result in increased selfing. Of the three orchards we studied, Hortlax had the highest clone number (62), but surprisingly, the greatest homozygosity excess in embryos. Thus, in this case the clone number did not seem to have much effect upon the genetic structure of the orchard embryos studied. However, the clone number given is the initial number. Thus it is possible that at the time of seed collection the number of surviving clones and ramets in Hortlax was smaller than initially. No inventory of the clone and ramet number was made at the time of seed collection. The differences between the seed orchards with regard to fixation indices in embryos were probably not due to the age of clones. Surprisingly, the youngest seed orchard Alnön had the smallest fixation indices in embryos. On the other hand, it is possible that an unusual design of the Hortlax seed orchard was partly responsible for increased homozygosity in the embryos from this orchard. Each block in Hortlax was surrounded by a 7 m high hedge. Most of the mating may have occured within, rather than between blocks, which could increase the rates of selfing.

Furthermore, it is also likely, that the genetic composition of the crops varied between the blocks. Subsequent pooling of the seed crops from different orchard blocks could result in a Wahlund effect, i.e. a surplus of homozygotes caused by variation of allelic frequencies between blocks. Another feature that distinguished Hortlax from the other two seed orchards is its northern location and clone origin. Many observations indicate better opportunities for panmixia in the south than in the north. In fact, the southernmost seed orchard Alnön had little homozygosity excess in embryos, in spite of the smaller clone number and slightly younger age. Clones from high latitudes are frequently moved to more southern areas where milder climatic conditions are believed to enhance flowering and seed production. Considering the excess homozygosity in the Hortlax orchard, it seems that the present transfers of clonal material may not be sufficient to ensure good genetic efficiency in all orchards. Better interpretation of our results requires data on rates of outcrossing in the populations. Without this information, it is difficult to judge whether selfing alone could account for the observed high fixation indices in embryos. Low fixation indices in young trees indicate that most of the inbred individuals have been removed. A similar shift in the genotypic structure between the embryo and the later life stages has been observed in earlier studies (Shaw and Allard 1982, Tigerstedt et al. 1982, Yazdani et al. 1985). As the young trees analysed in our study were only several years old, it appears that this elimination may occur rather early. Recently, Hiebert and Hamrick (1982) analysed genotypic structure in two-weeks old seedlings of Pinus longaeva. They found homozygote excess at only some of the loci studied while at others even heterozygote excess was occasionally found. Viability selection against inbred individuals may have already affected the genotypic structure in these seedlings. However, Tigerstedt et al. (1982) observed departures from panmictic genotype proportions in about 100 year old progeny of Scots pine. Unfortunately, we do not have current estimates of mortality in the field trials. Such data could resolve whether mortality after planting might account for the reduction of the fixation indices in the young plants or whether the elimination of inbreds took place before planting. According to observations made in Drängsmark and Raggsjö, from 6 to 15 % of the young plants had died by the year 1979. This is not sufficient to account fot the observed reduction in the fixation indices even if most of the mortality was against inbred homozygotes. However, later observations made in these trials indicate that much additional mortality had occured before our samples were collected in 1982. The results of these observations are being currently evaluated by the Institute of Forest Improvement at Sävar and should permit us to obtain better estimates later.

Acknowledgements. The authors wish to thank Dag Lindgren for helpful comments and critical discussion, Ola Rosvall, Torbjörn Lestander and Bengt Andersson for valuable information concerning the material analysed, Barbro Harbom, Maj-Lene Åman and Rigmor Hjalmarsson for excellent technical assistance, Stefan Löfmark for help in field collections, Od Nilsen for valuable comments on the design of the Hortlax orchard. This work was supported by a grant from the Cellulose Industries Council for Technical and Forest Research to A.E. Szmidt and O. Muona acknowledges support from the Ehrnrooth Foundation and the Academy of Finland. We also thank the participants of the meeting for useful comments on the study.

LITERATURE

Adams, W.T. 1981. Populations genetics and gene conservation in Pacific Northwest conifers. Proc. Conf. "Evolution Today" JCSEB II: 401-415.

Curie-Cohen, M. 1982. Estimates of inbreeding in a natural population: a comparison of sampling properties. Genetics 100: 339-358.

Ericsson, T. and Hadders, G. 1975. Odlingstest av tallplantagefrö. Institutet för Skogsförbättring Note No. 5: 1-4.

Ericsson, T. 1980. Odlingstest av tallplantagefrö i Norrland. Institutet för Skogsförbättring Note No. 7: 1-4.

Eriksson, G., Schelander, B. and Åkebrand, V. 1973. Inbreeding depression in an old experimental plantation of Picea abies. Hereditas 73: 185-194.

Friedman, S.T. and Adams, W.T. 1982. Genetic efficiency in loblolly pine seed orchards. Proc. 16th Southern Forest Tree Improvement Conf.: 213-220.

Harlan, J.R. 1975. Our vanishing genetic resources. Science 188: 618-621.

Hiebert, R.D. and Hamrick, J.L. 1983. Patterns and levels of genetic variation in great basin bristlecone pine Pinus longaeva. Evolution 57: 302-310.

Jonsson, A., Ekberg, I. and Eriksson, G. 1976. Flowering in a seed orchard of Pinus sylvestris L. Studia Forestalia Suecica 135: 1-38.

Kleinschmit, J. 1979. Limitations for restriction of the genetic variation. Silvae Genet. 28: 61-67.

Koski, V. 1973. On self-pollination, genetic load and subsequent inbreeding in some conifers. Commun. Inst. For. Fenn. 78: 1-42.

Libby, W.J. 1973. Domestication strategies for forest trees. Canad. J. For. Res. 3: 265-276.

Mejnartowicz, L. 1979. Genetic variation in some isoenzyme loci in Scots pine (Pinus sylvestris L.) populations. Arboretum Kórnickie 24: 91-103.

Moran, G.F., Bell, J.C. and Matheson, A.C. 1980. The genetic structure and levels of inbreeding in a Pinus radiata D. Don seed orchard. Silvae Genet. 29: 190-193.

Muona, O. and Szmidt, A.E. 1985. A multilocus study of natural populations of Scots pine. (this volume).

Müller-Starck, G. 1982a. Reproductive systems in conifer seed orchards. I. Mating probabilities in a seed orchard of Pinus sylvestris L. Silvae Genet. 31: 188-197.

Müller-Starck, G. 1982b. Sexually assymetric fertility selection and partial self-fertilization. 2. Clonal gametic contributions to the offspring of Scots pine seed orchard. Silva Fennica 16: 99-106.

Nagasaka, K. and Szmidt, A.E. 1984. Multilocus analysis of external pollen contamination of a Scots pine (Pinus sylvestris L.) seed orchard. Submitted, this volume.

O'Reilly, C.O., Parker, W.H. and Barker, S.E. 1982. Effect of pollination period and strobili number on random mating in a clonal seed orchard of Picea mariana. Silvae Genet. 31: 90-94.

Phillips, M.A. and Brown, A.H.D. 1977. Mating system and hybridity in Eucalyptus pauciflora. Aust. J. Biol. Sci. 30: 337-344.

Rudin, D. 1977. Leucine-amino-peptidases (LAP) from needles and macrogametophytes of P. sylvestris L. Inheritance of allozymes. Hereditas 85: 219-226.

Rudin, D. and Ekberg, I. 1978. Linkage studies in Pinus sylvestris L. using macrogametophyte allozymes. Silvae Genet. 27: 1-12.

Shaw, D.W. and Allard, R.W. 1982. Isozyme heterozygosity in adult and open-pollinated embryo samples of Douglas-fir. Silva Fennica 16: 115-121.

Skogsstatistisk Årsbok. 1980-1982. Sveriges officiella statistik, Skogstyrelsen, Jönjöping.

Smith, D.B. and Adams, W.T. 1983. Measuring pollen contamination in clonal seed orchards with the aid of genetic markers. Proc. 17th Southern Forest Tree Impr. Conf. Univ. of Georgia, Athens: 69-77.

Sokal, R.R. and Rohlf, R.F. 1981. Biometry. Freeman, San Francisco.

Szmidt, A.E. and Yazdani, R. 1983. Electrophoretic studies of genetic polymorphism of shikimate and 6-phosphoglucose dehydrogenases in Scots pine (Pinus sylvestris L.). Arboretum Kórnickie 28: in press.

Szmidt, A.E. and Yazdani, R. Inheritance of aconitase and phosphoglucose isomerase electrophoretic variants in Scots pine (Pinus sylvestris L.) tissues. (in prep.).

Tigerstedt, P.M.A., Rudin, D., Niemelä, T. and Tammisola, J. 1982. Competition and neighbouring effect in a naturally regenerating population of Scots pine. Silva Fennica 16: 122-129.

Yazdani, R. and Rudin, D. 1982. Inheritance of fluorescent esterase and β-galactosidase in haploid and diploid tissues of Pinus sylvestris L. Hereditas 96: 191-194.

Yazdani, R., Muona, O., Rudin, D. and Szmidt, A.E. 1985. Genetic structure of a Pinus sylvestris L. seed-tree stand and naturally regenerated understorey. Forest Science (in press).

GENETIC DIFFERENTIATION AMONG SCOTS PINE POPULATIONS FROM THE LOWLANDS AND THE MOUNTAINS IN POLAND*)

L. Mejnartowicz and F. Bergmann

Abstract

Genetic variation within and differentiation among nine Scots pine populations from the lowlands and the mountains in Poland were examined at nine gene loci coding for five enzyme systems involved in the energy and amino acid metabolism.

In general, the genetic variation in populations estimated as gene diversity appears to be higher than in other conifer species. However, lowland populations do not reveal a higher level of gene diversity than those from the mountains.
Based on estimates of genetic distance and a new measure of genetic differentiation, it was found that a considerable differentiation exists among lowland populations from different regions and between lowland and montane populations, whereas much lower differentiation was observed among populations from the mountains, thus roughly reflecting the different geographical distances. On the other hand, it was suspected that the low degree of differentiation found between the montane populations and the overall gene pool may indicate their relationship to former glacial refugia.

Introduction

Scots pine (*Pinus sylvestris* L.) is one of the coniferous tree species that has the widest range of distribution in Eurasia. In Poland, Scots pine is the most important forest forming tree species covering about 75% of the forest area, usually as pure pine stands.
In the northeastern part of this country, there also occur some of the economically best provenances of Europe (Giertych 1979).

In many studies with this tree species, geographic variation among different provenances could be demonstrated. For instance, Langlet (1959) using biometric methods has established clinal variation in metric traits such as seed weight, needle colour, time of bud set etc. Considerable genetic differentiation between latitudinally distant

*) This study was supported by funds MR II.16.3 from the Polish Academy of Sciences

populations and much lower variation between populations from different longitudes was found by Mejnartowicz (1979) on the basis of gene frequencies at three isozyme gene loci. Some genetic heterogeneity among Swedish Scots pine populations was observed by Rudin and co-workers (1974), and considerable geographic variation in isozyme patterns was found by Prus-Glowacki and Szweykowski (1977) und by Krzakowa (198o, 1982) among Polish Scots pine provenances.

The objective of our study was to compare the gene frequencies and gene diversity of Polish Scots pine populations from the mountains and the lowlands. It is assumed that pine stands growing at the vertical tree line of the species differ genetically from those growing in the lowlands, well within the natural range of the species. Earlier genetic studies with other conifers such as Norway spruce (Bergmann 1978), silver fir (Mejnartowicz 1979), and ponderosa pine (Mitton et. al. 198o) have shown that populations from different elevations may be very different genetically. To determine the gene frequencies of pine populations, nine isozyme gene loci, the products of which are mostly involved in the energy and amino acid metabolism of plants, were analyzed. If the genetic polymorphisms of these enzymes are functionally adaptive and related to climatic factors, a genetic differentiation between the montane and lowland populations may be assumed.

Material and Methods

This study included nine Scots pine populations of which four are located in the mountains, four in the lowlands and one is located in a sub-mountainous region on a peat-bog. It is suspected that in this latter location (Bór na Czerwonem) natural hybridization occurs between *Pinus sylvestris* and *Pinus mugo*. Some descriptions of the populations are given in Table 1, the geographical location of the populations is shown in a small map (Fig. 1).

Seed lots were collected from individual trees (15-2o per population) of six populations, whereas bulk seed collections were obtained from three populations. Haploid endosperms (macrogametophytes) of single seeds were used for isozyme analysis, so that a direct counting of allele frequencies was possible (Feret and Bergmann 1976, Mitton 1983).

Table 1: Geographic data of nine Scots pine populations from Poland

Name of population	altitude	district	abbreviation
mountain sites			
Dolina Koryciska Wielkie	95o m	Tatra mountains	Tatra
Macelowa mountain-top	8oo m	Pieniny "	Mac-top
Macelowa mountain-foot	6oo m	" "	Mac-foot
Zamkowa Mt. Muszyna	58o m	Beskidy "	Muszyna
sub-mountain site			
Bór na Czerwonem	45o m	spur of Beskidy	Bór-nC
lowland sites			
Kowalik	12o m	Masurian lakes	Kowa
Bolewice	8o m	Western Poland	Bole
Karczma Borowa	65 m	" "	Kar-Bo
Rychtal	15o m	Lower Silesia	Rych

Extracts from single endosperms containing soluble enzymes were analyzed by means of horizontal starch gel electrophoresis. Following the electrophoretic separation, the gel slabs were stained for five enzyme systems. Malate dehydrogenase (MDH, EC 1.1.1.37) and isocitrate dehydrogenase (NADP dependent IDH, EC 1.1.1.42) were resolved in a Tris-citrate buffer system (Siciliano and Shaw 1976); glutamate dehydrogenase (GDH, EC 1.4.1.2), glutamate oxaloacetate transaminase (GOT, EC 2.6.1.1), and formiate dehydrogenase (FDH, EC 1.2.1.2) were resolved in the Poulik system (Poulik 1957). The histochemical stains used for the visualization of these enzymes were described by Siciliano and Shaw (1976) and Tsay and Taylor (1978).

Gene frequencies were computed from the genotype numbers of trees. Each tree was genotyped using 6-9 endosperms per enzyme system. From bulk collections, gene frequencies were directly counted from zymograms of several hundred seed endosperms per sample. The gene frequency data were used to estimate the genic diversity by methods

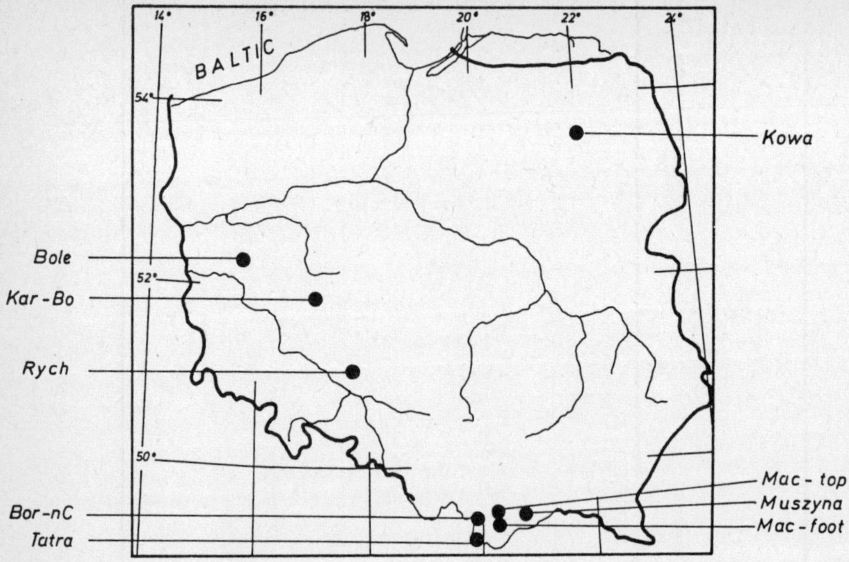

Figure 1. A map showing the location of the nine populations in Poland

published by Nei (1975) and by Gregorius (1978), to calculate the genetic distance defined by Gregorius (1974, 1984), and to measure the genetic differentiation by a method developed by Gregorius and Roberds (submitted for publication. Some features of this measure are explained by Gregorius in this volume).

Results

1) Variation and genetics of the enzyme systems

The genetic control of a variable activity zone in zymograms of an enzyme system was inferred from a 1:1 segregation ratio of isozyme variants found in a sample of haploid seed endosperms collected from a putative heterozygous mother tree (Feret and Bergmann 1976). If two (or more) zones of activity appear in the zymogram of an enzyme system, each zone was attributed to the action of a separate gene locus provided that the variation in one zone did not affect the variation or staining intensity of the other zone.

The glutamate oxaloacetate transaminase (GOT, identical to aspartate aminotransferase) system of Scots pine is controlled by three gene loci (GOT-A, GOT-B, and GOT-C), two of which are found to be polymorphic in the populations studied (Table 2). Five alleles were observed at the locus GOT-B, whereas even seven alleles appeared at GOT-C. Similar results were found by Müller-Starck (1982) and by Krzakowa (1982), while Rudin (1975) has identified two GOT loci and Chung (1981) four GOT loci in Scots pine. In the latter study, an enzyme band co-migrating with isozyme variants of GOT-C (double-banded variants) has probably been classified as a separate gene locus (GOT-D).

The glutamate dehydrogenase (GDH) system is controlled by one gene locus (GDH-A) as is the case in many other conifers. Chung (1981) has also identified only one GDH locus in Scots pine, whereas Krzakowa (1982) has observed three gene loci when resolving this enzyme in a Tris-citrate buffer system. In fact, three activity zones appear in Tris-citrate gels after staining for GDH. However, two of those zones can also be detected without adding the staining substrate, so that they are assumed to be other dehydrogenases. The same phenomenon can be observed in Tris-citrate gels stained for formiate dehydrogenase (FDH) which in fact is controlled by only one gene locus (FDH-A) in Scots pine. Both enzyme loci, GDH-A and FDH-A, are polymorphic in the pine populations investigated, exhibiting two and three alleles, respectively (Table 2).

The malate dehydrogenase (MDH) system is controlled by four gene loci, however, one invariant locus coding for an enzyme zone mostly masked by other MDH zones was omitted from the present study. The other three loci (MDH-A, MDH-B, and MDH-C) are largely polymorphic in the Scots pine populations. While MDH-A is found to have two alleles, four allelic types were generally observed at MDH-B and MDH-C (Table 2). However, two alleles at MDH-B (B_1 and B_3) could not be discovered in the populations studied. One of these alleles (B_1) was only detected in a population heavily stressed by industrial air pollution. Inheritance studies using full-sib families from German Scots pine seed orchards clearly show that the MDH system is spezified by four separate gene loci (Müller-Starck 1985, this volume).

The isocitrate dehydrogenase (IDH) system is presumably controlled by two gene loci. However, the slower migrating IDH zone in zymograms did

Population	GOT-A	GOT-B					GOT-C							FDH-A		
	A_1	B_1	B_2	B_3	B_4	B_5	C_1	C_2	C_3	C_4	C_5	C_6	C_7	A_1	A_2	A_3
Tatra	1.o	-	o.1o	o.27	o.63	-	o.13	o.2o	o.o7	o.23	o.37	-	-	-	o.2o	o.8o
Mac-top	1.o	-	o.o7	o.3o	o.53	o.1o	o.o7	o.23	-	o.o7	o.47	o.13	o.o3	-	o.15	o.85
Mac-foot	1.o	o.o3	o.16	o.22	o.59	-	-	o.38	o.o3	o.o9	o.44	o.o6	-	-	o.12	o.88
Muszyna	1.o	-	o.o9	o.34	o.53	o.o4	-	o.38	o.o3	-	o.28	o.31	-	-	o.1o	o.9o
Bor-nC	1.o	o.o8	o.17	o.25	o.5o	-	o.14	o.o3	o.o3	o.72	-	o.o3	o.o5	-	o.o8	o.92
Kowa	1.o	o.11	-	o.o1	o.88	-	-	o.2o	o.14	-	o.33	o.33	-	-	o.11	o.89
Bole	1.o	-	o.18	o.17	o.58	o.o7	o.15	o.12	o.o5	o.38	o.26	o.o3	o.o1	o.o3	o.15	o.82
Kar-Bo	1.o	-	o.17	o.18	o.64	o.o1	o.o6	o.58	o.o2	-	o.34	-	-	-	o.o8	o.92
Rych	1.o	o.o1	o.o8	o.31	o.5o	o.1o	o.o1	o.27	o.o5	o.o2	o.51	o.12	o.o2	o.o1	o.22	o.77

Population	IDH-A				MDH-A		MDH-B				MDH-C				GDH-A	
	A_1	A_2	A_3	A_4	A_1	A_2	B_2	B_4	B_5	B_6	C_1	C_2	C_3	C_4	A_1	A_2
Tatra	1.o	-	-	-	o.o3	o.97	o.87	-	o.13	-	o.23	o.4o	-	o.37	o.35	o.65
Mac-top	1.o	-	-	-	o.o4	o.96	o.82	-	o.18	-	o.o7	o.32	o.21	o.4o	o.22	o.78
Mac-foot	1.o	-	-	-	-	1.o	o.75	-	o.25	-	o.o3	o.38	o.o3	o.56	o.54	o.46
Muszyna	o.64	o.o1	o.o1	o.34	o.o7	o.93	o.67	-	o.33	-	o.o3	o.43	-	o.54	o.4o	o.6o
Bor-nC	1.o	-	-	-	o.o5	o.95	o.65	o.o5	o.3o	-	o.1o	o.5o	-	o.4o	o.32	o.68
Kowa	1.o	-	-	-	-	1.o	o.67	-	o.28	o.o5	-	o.72	-	o.28	o.34	o.66
Bole	o.93	-	o.o7	-	o.o2	o.98	o.66	o.o2	o.29	o.o3	o.o5	o.38	o.o8	o.49	o.38	o.62
Kar-Bo	1.o	-	-	-	o.13	o.87	o.61	o.o2	o.37	-	o.28	o.18	-	o.54	o.34	o.66
Rych	1.o	-	-	-	o.o5	o.95	o.82	o.o4	o.o9	o.o5	o.o8	o.27	o.65	-	o.23	o.77

Table 2. Allele frequncies at nine enzyme loci for nine Scots pine populations from Poland

not show clear banding patterns so that this zone was omitted from our study. The faster migrating zone with intensely stained IDH bands is controlled by IDH-A. Although this locus is monomorphic in most of the Scotspine populations, four alleles were detected in Muszyna and two alles in Bolewice (Table 2). The rare allele A_3 does not show enzyme activity in zymograms, so that it is regarded as a so-called null-allele.

2) Genetic variation within populations

Of the nine enzyme loci analyzed, six were polymorphic in all Scots pine populations studied (Table 2). While GOT-A was generally monomorphic, IDH-A was found to be monomorphic in seven and MDH-A only in two populations. Since the numbers and frequencies of alleles at most polymorphic loci vary from population to population (Table 2), the genetic variation averaged over all loci was expressed by a number of measures. They include P, the proportion of polymorphic loci in the population, and A/L, the mean number of alleles per locus based on polymorphic and monomorphic loci. The gene diversity of each population was measured by H_e (Nei 1975) and $v_{(p)}$ Gregorius 1978). However, H_e was not used as an estimate of heterozygosity because this assumes random mating and other population conditions often overlooked by researchers.

The values of the variation and diversity measures given in Table 3 show that there are great differences among populations. H_e, for instance, ranged from o.27 in Kowalik to o.38 in the Muszyna population. Based on all data of Table 3, Muszyna and Bolewice generally exhibit the highest degree of gene diversity which partly results from the IDH-A polymorphism detectable in these populations (Table 2). In contrast, Kowalik was found to have the lowest level of genetic variation (Table 3) which is caused by a reduced polymorphism at the MDH loci (Table 2). Depending on the greater sensitivity to allele numbers and frequencies (Bergmann and Gregorius 1979), the measure $v_{(p)}$ is proved to be more discriminatory than H_e (Table 3).

3) Genetic differentiation among populations

When comparing the gene frequencies of the nine populations, no pronounced differences between the montane and lowland populations could be established (Table 2). Only the frequencies of MDH-B_2 versus B_5

Table 3: Data of genetic variation and gene diversity for nine Scots pine populations estimated by different measures.

Name of population	P	A/L	H_e	$v_{(p)}$
Tatra	o.77	2.3	o.33	86.72
Mac-top	o.77	2.7	o.33	85.74
Mac-foot	o.66	2.4	o.32	58.11
Muszyna	o.88	2.7	o.38	147.47
Bór-nC	o.77	2.7	o.32	55.18
Kowa	o.66	2.1	o.27	31.69
Bole	o.88	3.2	o.37	155.75
Kar-Bo	o.77	2.4	o.33	61.43
Rych	o.77	3.1	o.33	6o.o6

were found to vary with the elevation of the population sites with the exception of Rychtal which, however, generally reveals extreme gene frequency distributions.

Based on the data given in Table 3, there is also no general difference in gene diversity between the montane and lowland populations. The mean values for A/L (2.5), H_e (0.34) and $v_{(p)}$ (94.5) averaged over the four montane populations are very similar to those averaged over the four lowland populations (A/L: 2.7, H_e: 0.33, $v_{(p)}$: 77.3). As a result, there was no trend of decreasing genetic variation with increasing elevation of population sites in Scots pine, as was reported for other tree species.

The estimates of genetic distance for all pairwise combinations of populations are presented in Table 4. Relatively high values were found for the distance among lowland populations and between lowland and montane populations, whereas much lower values were calculated for the distance among the montane population thus reflecting a greater genetic similarity. On the average, the highest values are found for combinations with Rychtal, a population from Lower Silesia.

In addition, the extent of genetic differentiation between each individual population and the whole collection was determined by another measure the characteristics of which are described in detail in a preceding paper of this volume (Gregorius 1985). Assuming equal population size for the nine Scots pine populations studied, the values

Table 4 : Estimates of genetic distance (d_o) for nine Scots pine populations based on data from nine enzyme loci

	Tatra	Mac-top	Mac-foot	Muszyny	Bor-nC	Kowa	Bole	Kar-Bo	Rych
Tatra	--								
Mac-top	0.100	--							
Mac-foot	0.115	0.110	--						
Muszyna	0.172	0.160	0.125	--					
Bor-nC	0.138	0.167	0.176	0.193	--				
Kowa	0.156	0.172	0.153	0.164	0.190	--			
Bole	0.103	0.132	0.114	0.153	0.142	0.168	--		
Kar-Bo	0.131	0.157	0.121	0.145	0.170	0.178	0.143	--	
Rych	0.156	0.080	0.185	0.221	0.236	0.220	0.200	0.216	--

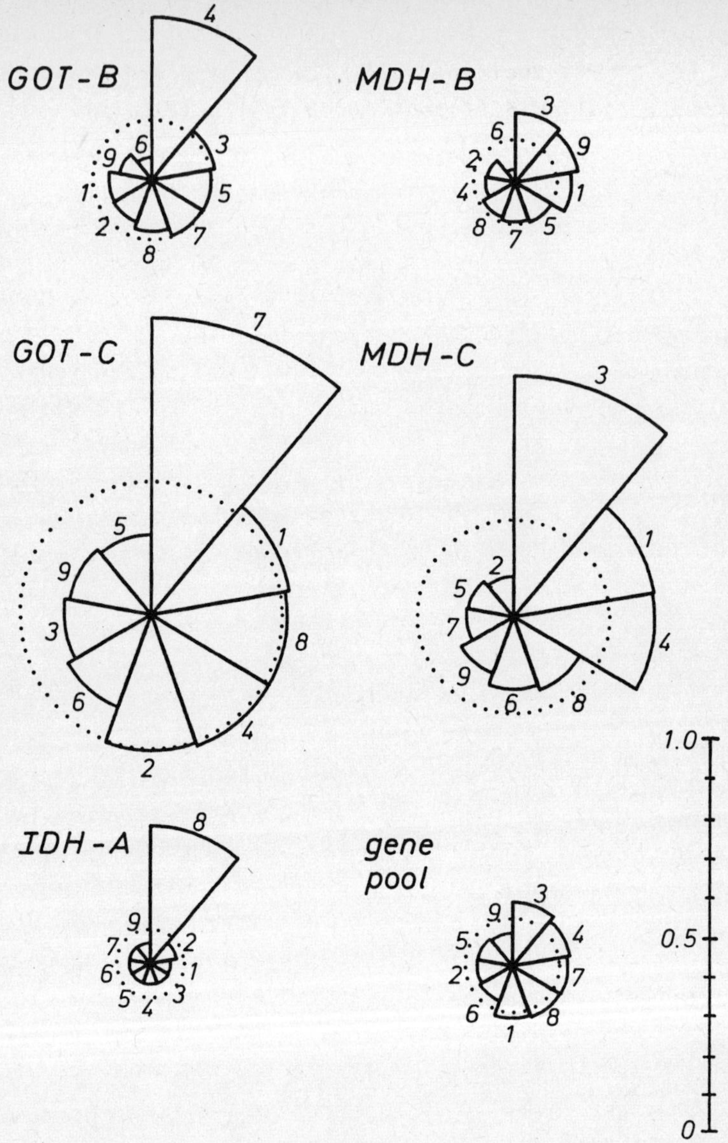

Figure 2. Differentiation snails for five enzyme loci and the corresponding gene pool. In each snail the dotted circle has a radius equal to the total level of genic differentiation. The solid sectors represent the contribution of individual populations to the total differentiation of the collection of populations.

for δ and D_j were computed using the gene frequency data of five out of the nine enzyme loci analyzed. The results are presented graphically by so-called differentiation snails (Fig. 2).

As can be seen by the different numbers, different populations are most strongly differentiated from the collection at different enzyme loci, so that the contribution of each population to the overall level of genic differentiation varies from locus to locus. Represented by solid sectors, Kowalik (4) at locus GOT-B, BoR-nC (7) at GOT-C, Muszyna (8) at IDH-A and Rychtal (3) at MDH-C have unique genetic structures strongly differing from those of the respective collection (Fig. 2). The differentiation snail für the gene pool comprising the data of the five loci reveals that Rychtal (3) is generally more differentiated from the other Scots pine populations thus resembling the data of the genetic distance (Table 4). On the other hand, Macelowa-top (5) and Tatra (9) show the lowest degree of differentiation from the collection indicating a genetic structure nearly representative for the Scots pine race in Poland.

Discussion

This study based on gene frequencies at nine enzyme loci shows that Scots pine populations in the Polish range contain a considerable amount of genic variation, which has already been indicated in a previous report (Mejnartowicz 1979). The estimates of genetic variation and gene diversity (H_e) are generally higher than those observed for other conifers (Mitton 1983), although the enzymes analyzed have polymeric structures and belong to the so-called group I enzymes (Gillespie and Langley 1974). For example, the gene loci coding for MDH and GDH which have often been found to be monomorphic exhibit great allelic variation in our Scots pine populations.

The effect of the altitudinal position on the level of genetic variation was evaluated by comparing the diversity data of populations from the mountains with those of populations from the lowlands. As shown in Table 3, there is no general trend towards decreased genetic variation with increased altitudinal position in the mountains as, for instance, was observed for Norway spruce (Bergmann and Gregorius 1979). A comparison of the populations from the top and the foot of Macelowa Mt. even reveals a somewhat higher gene diversity at the top of the mountain. Since both sites are covered with large, naturally regenerating populations, micro-environmental differences existing between the two sites were expected to give rise to genetic differences between these populations. Although there are some differences in allele frequencies at MDH-C and GDH-A (Table 2), a pronounced genetic differentiation could not be observed. This may be attributed to an effective gene flow via long-distance pollen dispersal and/or to a relatively low response of variation at the loci studied to selection caused by micro-climatic and edaphic heterogeneity.

Another result of this investigation was the inconsistent pattern of genetic differentiation among the nine Scots pine populations in Poland. As is seen from Fig. 2, the differentiation pattern changes considerably from locus to locus. Based on the differentiation snail for the gene pool, Rychtal (3) and Kowalik (4) reveal the highest degree of differentiation from the whole collection of populations. While Kowalik located in northeast Poland is geographically separated by a long distance from all other populations (Fig. 1), Rychtal located in Lower Silesia has probably been affected by air pollution stress.

On the other hand, Tatra (9) and Macelowa-top (5) show the lowest degree of differentiation, thus representing the genetic structure of the whole collection (Fig. 2). This result is in good agreement with the hypothesis that in Poland many plants including Scots pine have had their glacial refugia in the Pieninymountains from where they have spread in the postglacial period. Hence, the populations of Pieniny Mt. and neighborhoods are assumed to possess the original genetic structure of this forest tree species in Poland.

References

Bergmann, F. 1978. The allelic distribution at an acid phosphatase locus in Norway spruce (Picea abies) along similar climatic gradients. Theor. Appl. Genet. 52, 57-64.

Bergmann, F. and Gregorius, H.-R. 1979. Comparison of the genetic diversities of various populations of Norway spruce (Picea abies). In: Proc. Conf. Biochem. Genet. Forest Trees. D. Rudin (ed.), Umeå, Sweden, pp. 99-1o7.

Chung, M-S. 1981. Biochemical methods for determining population structure in Pinus Sylvestris L. Acta Forestalia Fennica 173, pp. 28.

Feret, P.P. and Bergmann, F. 1976. Gel electrophoresis of proteins and enzymes. In: Modern Methods in Forest Genetics, J.P. Miksche (ed.). Springer Verlag, New York, p. 49-77.

Giertych, M. 1979. Summary of results on Scots pine (Pinus sylvestris L.) heigh growth in IUFRO provenance experiments. Silvae Genetica 28, 136-152.

Gillespie, J.H. and Langley, C.H. 1974. A general model to account for enzyme variation in natural populations. Genetics 76, 837-884.

Gregorius, H.-R. 1974. Genetischer Abstand zwischen Populationen. I. Zur Konzeption der genetischen Abstandsmessung. Silvae Genetica 23, 22-27.

Gregorius, H.-R. 1978. The concept of genetic diversity and its formal relationship to heterozygosity and genetic distance. Bioscience 41, 253-271.

Gregorius, H.-R. 1984. A unique genetic distance. Biom. J. 26, 13-18.

Krzakowa, M. 198o. The most frequent genotype as indicator of interpopulational differentiation in Scots pine (Pinus sylvestris L.). In: Proc. IUFRO Symp. Scots Pine Forestry of the Future. Kórnik, Poland.

Krzakowa, M. 1982. Genetic differentiation of Scots pine populations. 1. Genotypes. Silva Fennica 16, 2oo-2o5.

Langlet, O. 1959. A cline or not a cline - a question of Scots pine. Silvae Genetica 8, 1-36.

Mejnartowicz, L. 1979. Genetic variation in some isoenzyme loci in Scots pine (Pinus sylvestris L.) populations. Arboretum Kórnickie 24, 91-1o4.

Mejnartowicz, L. 1984. Enzymatic investigations on tolorance in forest trees. In: Gaseous air pollutants and plant metabolism. M.J. Koziol and F.R. Whatley (eds.), Butterworths London, pp. 381-398.

Mitton, J.B., Sturgeon, K.B. and Davis, M.L. 198o. Genetic differentiation in ponderosa pine along a steep elevational gradient. Silvae Genetica 29, 1oo-1o3.

Mitton, J.B. 1983. Conifers. In: Isozymes in plant genetics and breeding. Part B S.D. Tanksley and T.J. Orton (eds.), Elsevier, Amsterdam, Oxford, New York, pp. 443-472.

Müller-Starck, G. 1982. Sexually asymmetric fertility selection and partial self-fertilization. 2. Clonal gametic contributions to the offspring of a Scots pine seed orchard. Silva Fennica 16, 99-1o6.

Nei, M. 1975. Molecular Population Genetics and Evolution, North-Holland Publ., Amsterdam-Oxford.

Poulik, M.D. 1957. Starch gel electrophoresis in a discontinuous system of buffers. Nature 18o, 1477.

Prus-Glowacki, W. and Sweykowski, J. 1977. Studies on isoenzyme variability in Scots pine (Pinus sylvestris) and mountain dwarf pine (Pinus mugo) populations. Bull. Sci. Amis. Poznań D 17, 15-27.

Rudin, D., Eriksson, G., Ekberg, I. and Rasmuson, M. 1974. Studies of allele frequencies and inbreeding in Scots pine populations by the aid of the isoenzyme technique. Silvae Genetica 23, 1o-13.

Rudin, D. 1975. Inheritance of glutamate-oxalate-transaminase (GOT) from needles and endosperms of Pinus sylvestris. Hereditas 8o, 296-3oo.

Siciliano, M.J. and Shaw, C.R. 1976. Separation and visualization of enzymes on gels. In: Chromatographic and Electrophoretic Techniques. I. Smith (ed.). W. Heinemann Med. Books Ltd., London, p. 185-2o9.

Tsay, R.C. and Taylor, I.E.P. 1978. Isoenzyme complexes as indicators of genetic diversity in white spruce. Picea glauca, in southern Ontario and the Yukon Territory. Formic, glutamic and lactic dehydrogenases and cationic peroxidases. Can. J. Botany 56, 8o-9o.

EFFECTS OF SELECTION PRESSURE BY SO_2 POLLUTION ON GENETIC STRUCTURES OF NORWAY SPRUCE (PICEA ABIES)

F. Bergmann and F. Scholz

Abstract

Air pollution is currently being shown to have many adverse effects on forest tree stands in Central Europe. However, ecological-genetic consequences of such a permanent stress factor were not considered up to now.

In order to obtain information about possible selection pressure caused by air pollution, numerous clones of Norway spruce were fumigated with SO_2 and subsequently ranked according to their visible damage. A comparison of a clone group of high sensitivity with one of low sensitivity revealed differences in genetic structure at four of eight enzyme gene loci studied. Surprisingly large difference in allele and genotype frequencies were found at one of these enzyme loci (G6PDH-A).

To demonstrate the possible effect of viability selection over the generations, the change in allele frequencies over several generations is predicted under certain simplifying assumptions concerning the fumigated clone material. It is shown that there is a very high probability of losing an allele at one or more of the four enzyme loci if air pollution stress persists.

Introduction

Damage to forest tree stands by air pollution is currently one of the most serious problems in Central Europe and, in recent years, also in Germany. Air pollution, a man-made environmental stress factor, causes many types of adverse effects on plant material, such as visible foliar injury, disturbance of biochemical pathways, growth reduction, and decreased seed crop, and can ultimately lead to the destruction of entire forest stands. Therefore, the forestry administration and the public demand elimination of the sources of pollution or, at least, drastic reduction of emissions of pollutants.

For population and ecological genetics, it is of primary importance to determine whether air pollution can exert, directly or indirectly, selection pressure in forest tree populations. The assumption of such a selection pressure is supported by several observations:

Many stands of forest tree species which show damages in highly polluted areas are found to include individual trees without any visible injury. Hence, it is reasonable to assume that this relative tolerance to air pollution is partly dependent on the genotype of the tree. Furthermore, in field experiments with different tree species, it could be shown that tree-to-tree variation in response to air pollution was largely due to genetic variance (for reviews, see Karnosky and Houston 1978, Mejnartowicz 1984).

Additionally, results from a few recent studies indicate that air pollution can affect pollen and egg cell production and fertilization in different ways (Houston and Dochinger 1977, Scholz et al. 1985).

If air pollution stress causes viability and/or fertility selection, then this stress can be dangerous for our forest tree populations because of its widespread and permanent occurrence. As a consequence a change in gene frequency towards extreme distributions and, ultimately, the irreversible loss of alleles at individual gene loci may result. Both events occurring in nearly the entire range of a species lead to a drastic decrease in genetic diversity, which is particularly disadvantageous for the adaptability of future generations of our long-lived forest tree species, which are forced to adapt to temporally and spatially varying environments (Hamrick 1979, Gregorius et al. 1979).

The objective of our investigation, therefore, was to establish to which degree and in what direction air pollution acts as a selective force. Since genetic changes due to selection effects can precisely be determined only by tracing gene and genotype frequencies, we used isozyme polymorphisms as gene markers. The only previous studies using isozyme polymorphisms to examine the effect of air pollution were done with Scots pine in Poland (Mejnartowicz 1983) and with Norway spruce (Scholz and Bergmann 1984).

Our first experiments were performed on collections of clones (cuttings) which were fumigated in open top chambers. It is maintained that in controlled fumigation experiments, phenotypic differences in air pollution tolerance largely result from genotypic differences, while in field stands environmental heterogeneity and competition additionally contribute to this phenotypic variation. Since conifers are generally more damaged by air pollution than broad-leaved tree species, we chose Norway spruce (Picea abies) which is one of the most sensitive tree species in

Germany. In addition, this tree species has been the object of numerous population and ecological genetic studies (for review, see Mitton 1983). In this study the genotype frequencies of more and less damaged clones were used to demonstrate the effect of viability selection.

Material and Methods

Young trees comprising various half-sib families from different Norway spruce provenances were repeatedly propagated by cuttings. This clonal material was fumigated with SO_2 in field chambers for two or three weeks until the phenotypic variation in damage ranged from "not visibly injured" to "total necrosis of current year needles" (Scholz, in prep.). From 123 clones fumigated and scored in this manner, the 30 most tolerant (subset S^-) and 31 most sensitive (subset S^+) were selected for a comparison of their genetic structure (Fig. 1).

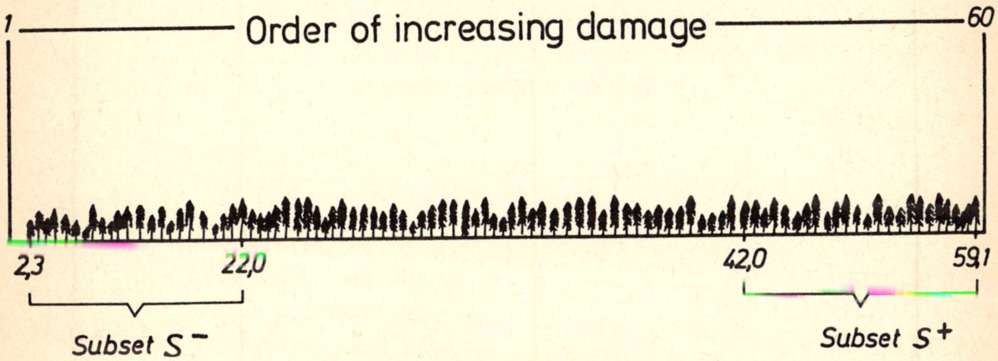

Figure 1. Schematic illustration of the division of the clones fumigated with SO_2 into the two subsets S^- (30 clones ranging from "no visible injury" to "low p.c. of damaged needles") and S^+ (31 clones ranging from "high p.c. of damaged needles" to "almost all needles damaged"). The numbersdesignate the ranking of the clones according to visible injury.

For isozyme analysis, meristem tissue of young vegetative buds collected from the non-fumigated ortets was extracted, and these extracts were then subjected to horizontal starch gel zone-electrophoresis. The following isozyme systems (and gene loci) were used: Glutamate oxaloacetate transaminase (GOT-B), glutamate dehydrogenase (GDH-A), malate dehydrogenase (MDH-C), isocitrate dehydrogenase (IDH-A, IDH-B), glucose-6-phosphate dehydrogenase (G6PDH-A), and phosphoglucomutase (PGM-A, PGM-B). Although some of these isozyme systems are controlled by more gene loci than given in brackets, we analyzed only those found to be polymorphic. Details of the electrophoretic procedures, staining recipes and identification of the genetic control were described elsewhere (Lundkvist 1979, Poulsen et al. 1983, Gregorius et al. 1984, Bergmann in prep.).

Results and Discussion

1) Genetic differences between the subsets

Assuming that the degree of visible injury of each plant after fumigation is largely dependent on its genotype, there must be differences in the genetic structure between subset S^- and subset S^+ (Fig. 1). To discover these genetic differences, we determined the allele and genotype frequencies at eight polymorphic enzyme loci for both subsets. In order to find possible relationships between allozyme variation and adaptive biochemical function, we studied only those loci, the enzymes of which have been found to be biochemically affected by air pollutants (Jäger and Klein 1980, Rabe and Kreeb 1980, Mejnartowicz 1984).

The resulting data compiled in Table 1 show that some genetic differences between the two subsets occur at three enzyme loci (GOT-B, MDH-C, PGM-B), while large and statistically significant differences were found at one locus (G6PDH-A). At the locus GOT-B, for instance, the genotype B_1B_2 is absent in subset S^- but appears with frequency of 0.10 in subset S^+, resulting in allele frequencies of 0.0 and 0.05, respectively. The allele frequencies at the MDH-C locus reveal that allele C_3 is more frequent in subset S^+ because the genotypes C_3C_3 and C_2C_3 occur predominantly in the more sensitive subset. Similar results are found at the locus PGM-B where the allele B_1 is more frequent in subset S^+ (Table 1).

Particularly striking are the genotype frequencies found at the locus G6PDH-A, where the heterozygote A_1A_2 is relatively rare and the homozygote A_1A_1 completely absent in subset S^-, while, in contrast, both genotypes are present with moderate frequencies in subset S^+ (Table 1). Accordingly, there is a significant difference in frequency of occurrence of allele A_1 between the two subsets. The data of the G6PDH system indicate a possible involvement of the gene products in the sensitivity of clones to SO_2 fumigation, which,

Table 1 : Genotype and allele frequencies at eight enzyme gene loci in the two subsets (S^+, S^-) of Norway spruce clones (after Scholz and Bergmann 1984)

Enzyme locus	genotype	allele	subset S^+	subset S^-	X^2-test
GOT-B	B_1B_2		0.10	-	n.s.
	B_2B_2		0.42	0.43	
	B_2B_3		0.26	0.47	
	B_3B_3		0.22	0.10	
		B_1	0.05	-	n.s.
		B_2	0.60	0.67	
		B_3	0.35	0.33	
GDH-A	A_1A_2		0.06	0.03	n.s.
	A_2A_2		0.94	0.97	
		A_1	0.03	0.02	n.s.
		A_2	0.97	0.98	
MDH-C	C_1C_2		-	0.10	n.s.
	C_2C_2		0.87	0.83	
	C_2C_0		-	0.03	
	C_2C_3		0.10	0.03	
	C_3C_3		0.03	-	
		C_1	-	0.05	n.s.
		C_2	0.92	0.91	
		C_3	0.08	0.02	
		C_0	-	0.02	
IDH-A	A_3A_3		0.81	0.83	n.s.
	A_3A_4		0.19	0.17	
		A_3	0.90	0.92	n.s.
		A_4	0.10	0.08	
IDH-B	B_2B_2		1.00	0.97	n.s.
	B_2B_3		-	0.03	
		B_2	1.00	0.98	n.s.
		B_3	-	0.02	
G6PDH-A	A_1A_1		0.03	-	$P < 0.05$
	A_1A_2		0.23	0.03	
	A_2A_2		0.74	0.97	
		A_1	0.15	0.02	$P < 0.01$
		A_2	0.85	0.98	
PGM-A	A_1A_2		0.03	0.03	n.s.
	A_2A_2		0.97	0.97	
		A_1	0.02	0.02	n.s.
		A_2	0.98	0.98	
PGM-B	B_1B_2		0.13	0.03	n.s.
	B_2B_2		0.87	0.97	
		B_1	0.07	0.02	n.s.
		B_2	0.93	0.98	

however, must be verified in additional studies.

There is some explanation for the putative relationship between G6PDH-A alleles and sensitivity when regarding biochemical pathways in plants. In several studies it could be shown that environmental stress such as air pollution leads to an increase of the pentose phosphate cycle at the expense of the glycolysis (ref. in Rabe and Kreeb 1980). Since G6PDH is the key enzyme of the pentose phosphate cycle, it may be suspected that trees carrying allele A_1 and its enzyme product are not capable of sufficiently intensifying the pentose phosphate cycle, and hence are less resistant to air pollution effects.

2) Effects of viability selection over generations

Genetic differences between two clone collections characterized by different degree of visible injury do not a priori imply a selection effect. If the action of the pollutant (here SO_2) is to cause a selective change in genetic structure, the most damaged plants must die prior to reproduction. Since the consequences of selection, however, should be demonstrated to emphasize the dangers arising from continuously acting air pollution, several simplifying assumptions concerning the clone material of the two subsets were made:

a) The collection of Norway spruce cuttings is regarded as a population reproducing over generations
b) All genotypes of subset S^- survive to the reproductive age, while all other genotypes die earlier (viability selection)
c) All surviving genotypes have the same reproduction rate (no fertility selection)
d) The selection pressure by air pollution is constant over generations
e) Mating occurs randomly.

Based on this simple selection model, the fitness values, i.e. the parameters of viability, were calculated for each genotype at the four enzyme loci exhibiting allele frequency differences between the two subsets (Table 1). Using these data the frequencies of alleles and genotypes were determined for several generations by computer simulation. The effect of selection is shown by the change of allele frequencies over generations (Fig. 2).

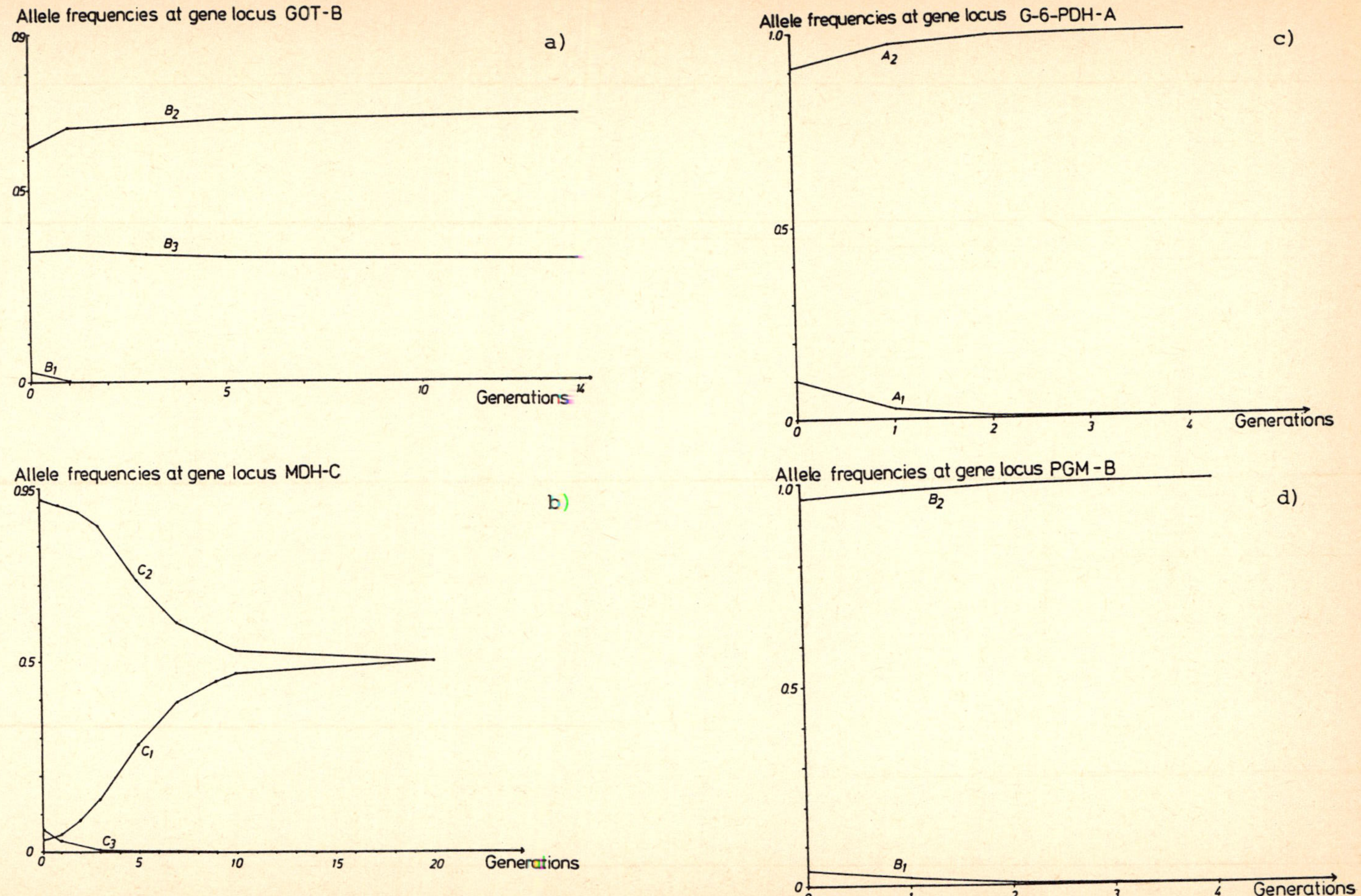

Figure 2. Changes in allele frequency at four enzyme loci over generation;
a) GOT-B, b) MDH-C, c) G6PDH-A, d) PGM-B

At the locus GOT-B, the allele B_1 disappears after one generation, while the frequencies of alleles B_2 and B_3 remain constant, resulting from the highest fitness of the heterozygote B_2B_3 (Fig. 2a). Similar frequency changes are seen for the locus MDH-C (Fig. 2b), where the allele C_3 approaches a frequency value of 0.0 after nine generations, so that it may well disappear due to the additional effect of drift. The polymorphism of alleles C_1 and C_2 is stable based on the highest fitness of the heterozygote. At the loci G6PDH-A and PGM-B, the alleles A_1 (Fig. 2c) and B_1 (Fig. 2d) approach very low frequencies after four generations and thus will be rapidly eliminated by even small drift effects. These two gene loci will most likely become monomorphic from an initially polymorphic state, resulting in a considerable loss of variability.

Concluding remarks

Although the effects of selection (Fig. 2) are based on several assumptions not in any case reflecting natural conditions in forest tree populations, they are yet important to demonstrate the possible consequences of permanent air pollution stress. In this simulation study, it is shown that the probability of losing an allele at the enzyme loci is very high after only a few generations. The consequence of such a gene loss is a decrease in genetic diversity resulting in a reduced adaptability in the next generation. On the other hand, we must bear in mind that these data were found for only a small proportion of spruce genes, and that probably by far more gene loci are involved in selective mechanisms caused by air pollution. Therefore, it is necessary to perform all known types of gene conservation, as long as most of the populations are not yet destroyed, so that the complete gene pool is present when reafforestation is possible after a considerable reduction of air pollution in Central Europe.

Acknowledgment

This study was supported by a grant from the Umweltbundesamt Berlin.

References

Gregorius, H.-R., Bergmann, F., Müller-Starck, G. and Hattemer, H.H. 1979. Genetische Implikationen waldbaulicher und züchterischer Maßnahmen. Allg. Forst- u. J.-Ztg. 150, 30-41.

Gregorius, H.-R., Hattemer, H.H. and Bergmann, G. 1984. Über Erreichtes und kaum Erreichbares bei der "Identifikation" forstlichen Vermehrungsguts. Allg. Forst- u. J.-Ztg. 155, 201-214.

Hamrick, J.L. 1979. Genetic variation and longevity. In: Topics in plant population biology. O. Solbrig, S. Jain, G. Johnson and P. Raven (eds.), Columbia Univ. Press, New York, pp. 84-113.

Houston, D.B. and Dochinger, L.S. 1977. Effects of ambient air pollution on cone, seed, and pollen characteristics in eastern white and red pines. Environ. Pollut. 12, 1-5.

Jäger, H.-J. and Klein, H. 1980. Biochemical and physiological effects of SO_2 on plants. Angew. Botanik 54, 337-348.

Karnosky, D.F. and Houston, D.B. 1978. Genetics of air pollution tolerance of trees in the northeastern United States. In: Proc. 26th Northeastern Forest Tree Improv. Conf., Pennsylv. State Univ., pp. 161-171.

Lundkvist, K. 1979. Allozyme frequency distributions in four Swedish populations of Norway spruce (Picea abies). Hereditas 90, 127-143.

Mejnartowicz, L. 1983. Changes in genetic structure of Scots pine (Pinus silvestris L.) population affected by industrial emissions of fluoride and sulphur dioxide. Genetica Polonica 24, 41-50.

Mejnartowicz, L. 1984. Enzymatic investigations on tolerance in forest trees. In: Gaseous air pollutants and plant metabolism. M.J. Koziol and F.R. Whatley (eds.), Butterworths London, pp. 381-398.

Mitton, J.B. 1983. Conifers. In: Isozymes in plant genetics and breeding. Part B. S.D. Tanksley and T.J. Orton (eds.), Elsevier, Amsterdam, Oxford, New York, pp. 443-472.

Poulsen, H.D., Simonsen, V. and Wellendorf, H. 1983. The inheritance of six isoenzymes in Norway spruce (Picea abies L. Karst.). Forest Tree Improv. 16, 12-33.

Rabe, R. and Kreeb, K.-H. 1980. Wirkungen von SO_2 auf die Enzymaktivität in Pflanzenblättern. Z. Pflanzenphysiol. 97, 215-226.

Scholz, F. and Bergmann, F. 1984. Selection pressure by air pollution as studied by isozyme-gene-systems in Norway spruce exposed to sulphur dioxide. Silvae Genetica 33, 238-241.

Scholz, F., Vornweg, A. and Stephan, B.R. 1985. Wirkungen von Luftverunreinigungen auf die Pollenkeimung von Waldbäumen. Forstarchiv 56, in press.

Measurement of genetic differentiation in plant populations

Hans-Rolf Gregorius

Abstract

It is argued that the studies of genetic differentiation within plant populations should explicitly take into account three major stages in the life cycle: adult plants, seed before dispersal, and seed after dispersal. The simultaneous consideration of these three stages is necessary in order to evaluate the effectiveness of the three main causes of differentiation, namely, non-random pollination and fertilization of the ovules, limited seed dispersal, and locally differential selection. In order to experimentally analyze genetic characteristics of adaptive strategies of plant populations, it is suggested that the association between genetic diversity and the combination of genetic differentiation at the three stages be considered.

The shortcomings of Wright's F_{ST} measure for population differentiation are pointed out, and a new measure is proposed which consistently combines the levels of differentiation for the individual subpopulations with the total level of population differentiation. The use of this measure is demonstrated with the help of two experimental data sets, one referring to differentiation in the mating system of pines (differentiation within populations), and the other concerned with differentiation among populations of wild barley in Israel (differentiation between populations). Appropriate data sets allowing simultaneous anlysis of the three major life cycle stages do not yet seem to have been obtained.

Introduction

Genetic differentiation of populations is normally viewed in connection with characteristics of habitats. Hence, the criteria for population subdivision are mainly of a spatial or an ecological nature or are a combination of these. The objective consists in finding genetic differences among the subpopulations or demes which correspond to the particular pattern of subdivision, and to relate such differences to the criteria for subdivision and to various components of the reproductive system of the population.

Criteria for subdivision which refer to characteristics of individuals rather than habitats (phenotypic, demographic, etc.) may not be admissible in studies of population differentiation, since these criteria may be genetically correlated and would, therefore, lead to tautological conclusions.

The major causes of differentiation in plant populations are non-random mating (limited pollen dispersal, incompatibilities, assortative mating caused by pollination vectors or asynchronous flowering, etc.), migration limited by low seed dispersal, and locally differential selection. The latter cause deserves particular attention in plants as compared with most animals, since large seed production allows for strong selection without endangering population survival, and individual immobility may thus enhance marked local selective differentiation during the phase between the seed and the adult stage. However, this tendency may be counteracted by the other two causes of differentiation if pollen or seed dispersal occurs over wide ranges. Consequently, the measurement of genetic differentiation within plant populations should explicitly account for the stage to which it is applied, since different stages may show almost reversed patterns of differentiation. Moreover, these stages should be directly related to the above three causes of differentiation.

A problem arises with the measurement of differentiation at the seed stage, since it reflects the effects of pollen and seed dispersal, i.e. the effects of the mating and migration system. In order to distinguish between these effects for a particular pattern of habitat subdivision it is necessary to consider the seed production before and after dispersal. If the seeds are collected from the mother plants, the corresponding differentiation pattern reflects the local adaptations of the adult plants via the ovules and effects due to differential fertilization by the pollen. In case it is possible to distinguish between the male and female contributions to a seed (e.g. with the help of the haploid endosperm in gymnosperms), the degree of differentiation which is solely due to the mating system can be computed by considering gene frequencies among the male contributions only. For this purpose, the population of seed bearing plants is subdivided into subpopulations (demes). The criterion for subdivision is determined by the trait with respect to which the mating system is to be evaluated. Hence, each deme may consist of a single plant if differential pollination and fertilization of individual plants is to be investigated. Other criteria, such as ecological, phenotypic or genetic, will generally lead to demes each comprising many individuals. The gene frequencies have to be determined for each deme within the pollen contributions to the seed production of this deme. If these frequencies are the same for all demes, the absence of mating preferences with respect to the locus and the deme structure can be assumed. Otherwise, differentiation due to the mating system can be postulated.

If the male and female contributions cannot be distinguished, it is still possible to estimate gene frequencies among the male contributions provided the same diploid locus can be identified in the adult and seed stage (e.g. ontogenetically stable gene expression for enzymes), and regular segregation can be assumed among the female contributions. In this case, a seed sample taken from a heterozygous mother plant contains, on the average, both alleles with equal frequency among the female contributions. Hence, the frequency p_i'' of the i-th allele among the female contributions to the seed production of a deme results from taking the weighted average of the frequency of this allele in the single mother plants. The weights are given by the sizes of the samples taken from the seed production of each plant (these sample sizes should be proportional to the actual seed output of each plant), and the allele frequency in each mother plant may take one of the values 0, 1/2 or 1, according to whether the plant does not contain the i-th allele, is heterozygous, or is homozygous for this allele, respectively. Since the frequency p_i of the i-th allele in the seed production of a deme is known, and since

$$p_i = 1/2 \cdot (p_i'' + p_i'),$$

where p_i' is the allele frequency among the female contributions, it follows that p_i' can be estimated as

$$p_i' = 2 \cdot p_i - p_i''.$$

In order to arrive at a more reliable picture of differentiation within a plant population, it is, according to the above remarks, advisable to compute degrees of differentiation at the three stages,i.e adult plants, seed before dispersal and seed after dispersal. The comparison of the resulting values allows conclusions to be drawn about the effectiveness of the corresponding three causes of differentiation. For example, high differentiation among the adult plants may be reduced to some extent among the seeds before dispersal because of random fertilization, and this amount may even vanish completely among the seed after dispersal in case the seeds are distributed at random over the habitat. There is some intuitive appeal to the idea that such a decreasing tendency in differentiation could be the rule. Exceptions to this rule, if they were observed, would hint at the existence of very special systems of mating or mechanisms of seed dispersal. This problem appears to have not yet been studied experimentally.

Moreover, the amount of genetic variability which can be maintained in a population critically depends on the local selective differences together with the amount of gene flow between the localities. This suggests that the association between genetic diversity and the combination of genetic differentiation at the three stages be considered in order to gain more insight into the adaptive principles characteristic of plant populations. If, for example, it should turn out that high genetic diversity is always accompanied by high genetic differentiation among the adult plants but low differentiation among the seed after dispersal, this would indicate a mechanism for the maintenance of genetic polymorphisms which acts through a cyclical switch between diversifying local adaptation and homogenizing migration. Herewith, it may make an important difference whether the homogenization is mainly due to wide range dispersal of pollen or of seed (Gregorius and Namkoong 1983, Namkoong and Gregorius 1985). These authors have demonstrated that, under the assumptions of their models, local selective differences are more effective in protecting genetic polymorphisms if seed dispersal is low and pollen dispersal occurs over a wide range. Hence, it may be valuable to obtain information about differentiation both before and after seed dispersal.

An appropriate measure of differentiation

The standard measure of genetic differentiation within and between populations is Wright's F_{ST} (cf. e.g. Wright 1978) which is identical to Nei's G_{ST} (Nei 1973). This measure was initially developed by Wright as an index of genetic fixation for his 'island model' of population subdivision, and it was later re-interpreted as a measure of differentiation. As the author himself pointed out, this re-interpretation is subject to some shortcomings which result from the original purpose of the index. In particular, F_{ST} assumes its maximum value of 1 only if all demes are genetically fixed, thus including the special case where several demes are fixed for the same allele, which fundamentally contradicts the concept of a completely differentiated population. Complete differentiation should be realized only if no pair of demes has alleles in common, which does not exclude the case where some or even all of the demes are polymorphic. Moreover, it is not quite clear whether F_{ST} has any meaning if differentiation is to be considered at the genotypic rather than the genic level.

An additional difficulty arises if the focus is on evaluating the local, i.e. deme specific, contributions to population differentiation. This problem occurs, for example, in studies involving the comparison of marginal and central parts of a population or parts distributed along an environmental gradient. Such situations are normally treated by considering the average genetic distance of a deme to the remaining demes (cf. e.g. Nevo et al. 1979, or Linhart et al. 1981). Herewith, in most cases Nei's genetic distance (Nei 1972) is applied, a measure which is known to increase indefinitely as a pair of populations becomes genetically more distinct. Hence, not only is it difficult to state what distances should be considered as large, but also there is no correspondence between deme specific differentiation and the total population differentiation as measured by F_{ST}.

In order to overcome these difficulties and arrive at a consistent construction of a measure of differentiation, it is necessary to set up a minimal number of conditions which such a measure should be expected to fulfill. Among these are that this measure should be equally applicable to all types of genetic frequencies (including allelic, multilocus genic, gametic and genotypic frequencies).Secondly, the values taken on by the measure should directly reflect the genetic differences among the demes so that 0 signifies the absence of differences and 1 signifies complete differentiation (i.e no two demes have any genetic types in common which exist in the population).

Furthermore, as was recently argued (Gregorius and Roberds 1984), a measure which reflects some of the most basic, intuitive ideas of differentiation should be built upon the genetic differences between each deme and its population complement. Such a difference is most appropriately specified as the proportion of genetic elements (alleles, genes at multiple loci, gametes, genotypes) by which a deme differs from the remainder of the population in type. This proportion is the amount of genetic differentiation of the deme, and it was shown by these authors that for the j-th deme it attains the representation

$$D_j = .5\cdot\sum_i |p_i(j) - \bar{p}_i(j)|,$$

where $p_i(j)$ is the relative frequency of the i-th genetic type in the j-th deme, $\sum_i p_i(j)=1$, and $\bar{p}_i(j)$ is the relative frequency of this type in the remainder of the population. The D_j's are in fact measures of genetic distance (Gregorius 1974, 1984) and are thus 0 or 1 if the j-th deme is genetically identical to or completely different from the remainder of the population, respectively. Note that D_j is not identical to the average genetic distance of the j-th deme from the remaining demes, but rather replaces this average as a tool for evaluating differentiation along spatial or environmental (ecological) gradients. Hence, clines of deme differentiation along such gradients should be described in terms of the D_j's.

The sum of the single deme differentiations expressed as the proportion among all genetic elements present in the population (population size) leads to a consistently defined measure of total population differentiation which is termed the level, δ, of differentiation among demes (subpopulations). Since the demes may vary in size, each deme must be weighted by c_j, the proportion of genetic elements present in the j-th deme ($\sum c_j = 1$), so that

$$\delta = \sum_j c_j \cdot D_j.$$

Thus, δ is the proportion of genetic elements by which the demes differ from their respective complements in type.

These measures can be applied to all types of genetic frequencies, as required, provided that the sets of genetic entities for which their frequencies have to be determined are properly specified. This becomes particularly relevant if gene frequencies are obtained separately for several loci, and it requires one to distinguish between genic, gametic and genotypic differentiation. The relationships between these measures and their population genetic significance is discussed in the above-cited paper by Gregorius and Roberds (1984). There it is also illustrated that, in particular, comparisons between differentiatation at the genotypic and the genic level can reveal interesting aspects of locally differential adaptations.

Application to experimental data sets

As is indicated at the beginning, there seem to exist no data sets which allow study of stage specific genetic differentiation in plants for all three relevant stages simultaneously. Among the experimental studies which treat the single stages separately, those concerned with seed before dispersal, and thus with aspects of the mating system, are very rare. In fact, only one publication appears to exist (Müller-Starck 1981) which provides information on gene frequencies that can be used to calculate indices of differentiation among the male contributions. This author investigated mating probabilities in a clonal seed orchard of Pinus sylvestris L. with the help of an enzyme locus. The clonal replicates in this orchard were distributed uniformly over the growing area, so that deviations from random mating resulting from preferential pollination among neighbouring trees cannot be detected at the locus studied . Hence, any genetic differences among the pollen contributions to the seed production of the single trees should be primarily due to factors such as increased self pollination, asynchronous flowering, incompatibilities, etc.

Since allele frequencies among the pollen contributions were obtained for the seed production of single trees, where the sample sizes were all identical (192 seeds), the seed production of each tree is considered as a deme, and all demes have the same size. The 14 trees investigated can be considered as a sample of all 'demes' present in the orchard. Table 1 contains the D_j and δ values for two flowering periods, 1974 and 1976. The deme numbers correspond to the tree numbers used by the author.

Table 1

Deme No.	88	174	185	177	229	69	34	134	157	76	149	126	179	220	δ (%)
D_j (%)															
1974	8.1	5.0	7.0	7.6	2.6	3.6	2.0	4.2	2.5	1.6	7.6	3.3	3.3	4.8	4.5
1976	3.4	2.2	13.0	5.7	7.2	8.3	3.9	3.8	4.6	3.8	3.5	4.5	0.5	5.5	5.0

Considering the fact that the population of adult plants was the same in both flowering periods, it is not surprising that the levels of differentiation, δ, are also almost the same. The pattern of single deme differentiation, however, is less easy to explain, particularly because some demes show relatively high differentiation in one year but low differentiation in the other (e.g. demes 88 and 69). This indicates that the ovules of the same tree are randomly fertilized in one but not in the other flowering period, at least with respect to the alleles at the locus under consideration. Deme 185 is an exception in that it shows high, above average differentiation in both years, particularly in 1976 (13%). The same applies to deme 177 to a lesser degree. This observation suggests that these trees may be worthy of further investigation. However, although some of the deme differentiations are statistically significant, the total level δ of differentiation is probably too small to justify more detailed interpretation. Moreover, the sampling procedure did not account for possible differences in the amount of seed production among the mother trees, a factor which might also affect the pattern of differentiation.

Apparently, the majority of experimental studies of population differentiation are concerned with genetic differences between rather than within populations. The present measures δ and D_j can also be applied to such situations where the demes or subpopulations are considered as more or less closed populations. A possible ambiguity may, however, occur in connection with the definition of genetic frequencies in the 'population complements', since these frequencies depend on the sizes of the complementary populations. If each population in the collection is viewed to be of equal significance, irrespective of its size, it is appropriate to attribute equal weights to all of them (c_j=1/N where N= number of populations in the collection and j=1,...,N). Otherwise, if small populations are considered to contribute less to the differentiation of the collection than large ones, sizes have to be taken into account as weights for the relative frequencies in each of the populations.

Probably because of difficulties in estimating population sizes, most investigations of genetic differentiation in collections of populations are evaluated under the assumption of

Table 2

	Loci					
	Acph-2	Acph-3	Adh-1	Est-1	Est-2	gene pool
D_1	100	73.0	14.3	6.7	98.2	58.5
D_2	25.9	72.0	14.3	5.0	91.8	41.8
D_3	54.5	28.4	21.8	6.7	63.8	35.1
D_4	45.3	28.8	20.8	6.7	74.0	35.1
D_5	51.8	45.0	91.3	16.9	63.7	53.5
D_6	30.6	20.1	37.4	11.2	54.1	30.7
D_7	29.7	48.1	91.3	20.0	77.2	53.3
D_8	17.6	58.5	14.3	6.7	66.0	32.6
D_9	86.6	62.6	14.3	6.7	92.3	52.6
D_{10}	64.9	59.5	8.3	6.7	80.9	44.1
D_{11}	52.4	50.1	14.3	69.5	85.4	54.4
D_{12}	24.1	30.9	12.3	6.7	64.1	27.6
D_{13}	79.3	44.0	14.3	10.4	72.1	44.0
D_{14}	59.1	42.9	52.0	6.7	76.4	47.4
D_{15}	46.1	21.0	14.3	8.6	53.1	28.6
D_{16}	53.7	22.6	14.3	6.2	72.4	33.8
D_{17}	60.1	57.4	14.3	5.0	78.9	43.1
D_{18}	49.7	59.5	14.3	6.7	65.2	39.1
D_{19}	55.9	29.8	14.3	6.7	66.4	34.6
D_{20}	27.7	18.0	14.3	6.7	59.0	25.2
D_{21}	21.9	21.1	14.3	6.7	66.7	26.2
D_{22}	45.8	22.6	14.3	6.7	60.8	30.0
D_{23}	45.3	19.1	14.3	6.7	78.7	32.8
D_{24}	54.2	36.7	14.3	6.7	53.4	33.1
D_{25}	33.4	25.7	14.3	6.2	44.1	24.7
D_{26}	71.1	68.8	14.3	6.7	67.3	45.6
D_{27}	72.3	41.9	14.3	6.7	67.1	40.5
D_{28}	14.7	35.7	14.3	18.8	52.4	27.2
δ	49.1	40.8	22.2	10.4	69.4	38.4

equal representation of all members of the collection. This also applies to the data set reported and analyzed by Nevo et al. (1979). The authors studied 28 populations of wild barley, Hordeum spontaneum, which were distributed across Israel. Genetic structures were identified at 28 enzyme loci, of which five (Acph-2, Acph-3, Adh-1, Est-1, Est-2) are selected here for the computation of genic δ and D_j values. The results (stated in %) are listed in Table 2, including the differentiation indices for the gene pool formed by the aforementioned 5 loci.

The levels of differentiation between populations are surprisingly high. Even at the locus with the lowest level (Est-1) the populations still differ from their complements in 10.4% of their genes. Moreover, population 2, for example, is almost representative of the whole collection (D_2=5%) at this locus, while population 11 is far away from being representative with its 69.5% differentiation. On the other hand, the Est-2 locus consistently shows high differentiation for all populations with a minimum of 44.1% for population 25. This suggests that the allelic variation at the Est-2 locus contains considerably more adaptive sensitivity to the environmental differences existing among the populations than the Est-1 locus. The other loci range between these extremes. This aspect can be demonstrated graphically with the help of the 'differentiation snails' displayed in the following figures. In each such snail the dotted circle has a radius equal to the genic level of differentiation (δ). The solid sectors represent the contributions of the single populations to the total differentiation of the collection of populations. The radii of the sectors are equal to the differentiation levels of the individual populations (D_j's) and the angles of the sectors represent the population weights (c_j's) which are all identical in the present case (c_j=1/28). The populations are arranged such that the D_j's appear in decreasing order, which yields the typical snail form.

The figures reveal additional interesting differences among the differentiation patterns of the individual loci. Evidently, the ranking of population differentiation may change considerably from locus to locus; i.e certain populations are differentiated strongly at some loci and weakly at others. The loci with lower levels of differentiation (Adh-1, Est-1) show the most drastic decline from the population with the highest degree of differentiation to the population with the lowest. This also explains the fact that only very few populations show above average differentiation at these loci.

These aspects are blurred in the gene pool representation. In particular, the important observation that some loci seem to have a greater capacity to react adaptively to the environmental differences between the populations than others cannot be inferred from the pattern of gene pool differentiation alone. Hence, it might be advisable in some situations not to combine the single locus observations into one measurement but rather to analyze each locus separately.

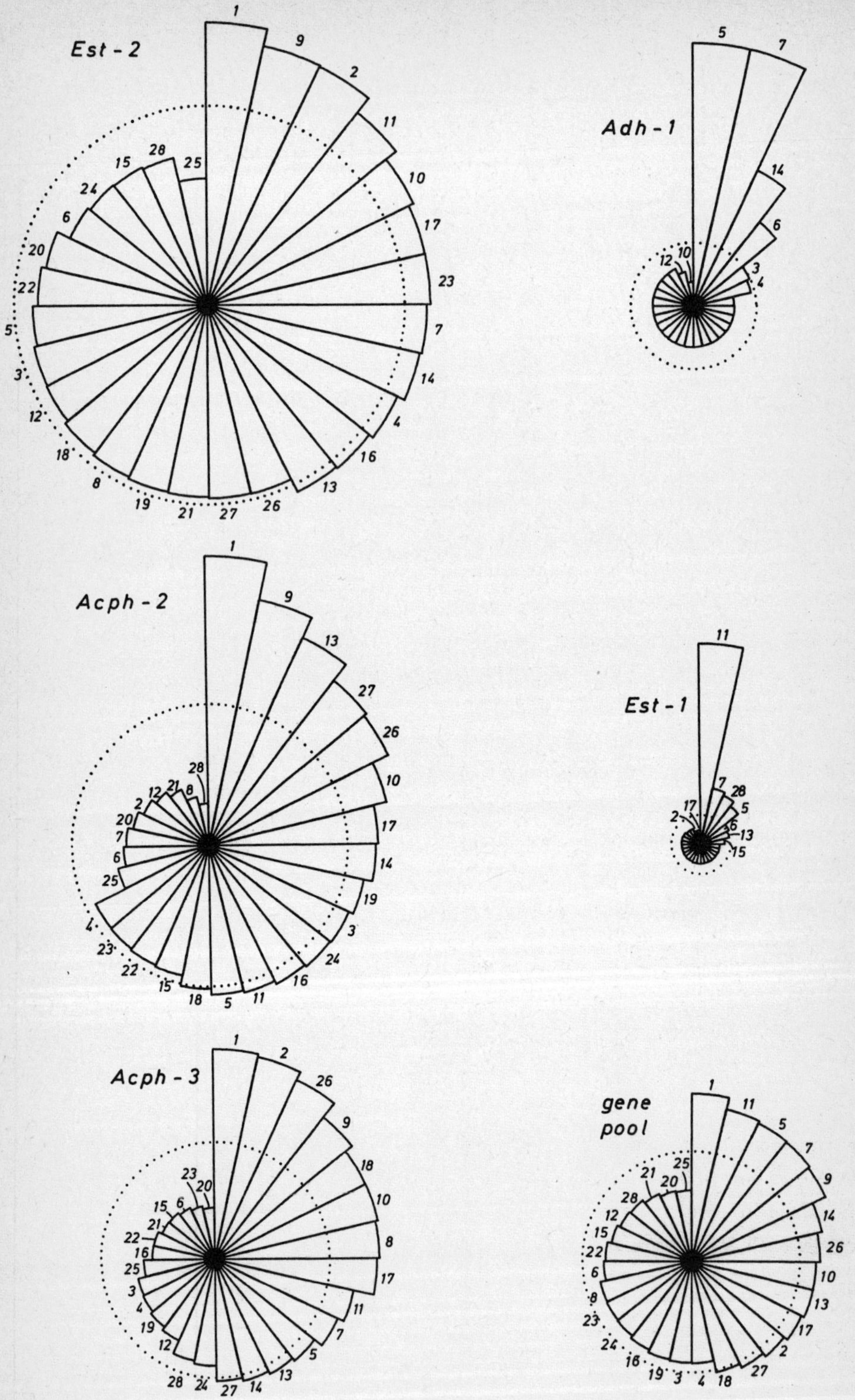
Est - 2
Adh - 1
Acph - 2
Est - 1
Acph - 3
gene pool

References

Gregorius, H.-R. 1974: Genetischer Abstand zwischen Populationen. I. Zur Konzeption der genetischen Abstandsmessung. Silvae Genetica 23: 22-27

Gregorius, H.-R. 1984: A unique genetic distance. Biometrical J. 26: 13-18

Gregorius, H.-R., G. Namkoong 1983: Conditions for protected polymorphisms in subdivided plant populations. 1. Uniform pollen dispersal. Theoretical Population Biology 24(3): 252-267

Gregorius, H.-R., J.H. Roberds 1984: Measurement of genetical differentiation among subpopulations. Submitted for publication.

Linhart, Y.B., J.B. Mitton, K.B. Sturgeon, M.L. Davis 1981: Genetic variation in space and time in a population of ponderosa pine. Heredity 46: 407-426

Müller-Starck, G. 1982: Reproductive systems in conifer seed orchards. I. Mating probabilities in a seed orchard of *Pinus sylvestris* L.. Silvae Genetica 31: 188-197

Namkoong, G., H.-R. Gregorius 1985: Conditions for protected polymorphisms in subdivided plant populations. 2. Seed vs. pollen migration. Amer. Natur., in press.

Nei, M. 1972: Genetic distance between populations. Amer. Natur. 106:283-292

Nei, M. 1973: Analysis of gene diversity in subdivided populations. Proc. Nat. Acad. Sci. USA 70(12): 3321-3323

Nevo, E., D. Zohary, A.H.D. Brown, M. Haber 1979: Genetic diversity and environmental associations of wild barley, *Hordeum spontaneum*, in Israel. Evolution 33(3): 815-833

Wright, S. 1978: Evolution and the Genetics of Populations. Vol.4. The University of Chicago Press.

LIST OF CONTRIBUTORS

BARRETT, S.C.H. : Department of Botany, University of Toronto, Toronto, Ontario. M5S 1A1, Canada

BERGMANN, F. : Abteilung für Forstgenetik und Forstpflanzenzüchtung, Universität Göttingen, Büsgenweg 2, D-3400 Göttingen, Federal Republic of Germany

BROWN, A.H.D. : CSIRO Division of Plant Industry, G.P.O. Box 1600, Canberra, A.C.T. 2601. Australia

BROWN, I.R. : University of Aberdeen, Department of Forestry, St. Machar Drive, Aberdeen, United Kingdom. AB9 2UU

CHARLESWORTH, B. : The University of Chicago, Department of Biology, 1103 East 57th Street, Chicago, Illinois 60637, USA

CHELIAK, W.M. : Petawawa National Forestry Institute, Chalk River, Ontario, KOJ 1JO Canada

ERNST, W.H.O. : Department of Ecology and Ecotoxicology, Biological Laboratories, The Free University of Amsterdam, P.O. Box 7161, NL-1007 MC Amsterdam, The Netherlands

GREGORIUS, H.-R. : Abteilung für Forstgenetik und Forstpflanzenzüchtung, Universität Göttingen, Büsgenweg 2, D-3400 Göttingen, Federal Republic of Germany

HAMRICK, J.L. : Department of Botany, University of Kansas, Lawrence, Kansas 66045, USA

HEDRICK, P.W. : Department of Biology, Division of Biological Sciences, University of Kansas, Lawrence, Kansas 66045, USA

LINDGREN, D. : Department of Forest Genetics and Plant Physiology, The Swedish University of Agricultural Sciences, S-901 83 Umeå, Sweden

MEJNARTOWICZ, L. : Polish Academy of Sciences, Institute of Dendrology, PL-63-120 Kórnik, Poland

MORAN, G.F. : CSIRO Division of Forest Research, P.O. Box 4008, Queen Victoria Terrace, A.C.T. 2600. Australia

MÜLLER-STARCK, G. : Abt. für Forstgenetik und Forstpflanzenzüchtung, Universität Göttingen, Büsgenweg 2, D-3400 Göttingen, Federal Republic of Germany

MUONA, O. : Department of Genetics, University of Oulu, SF-90570 Oulu, Finland

NAGASAKA, K. : Department of Forest Genetics and Plant Physiology, The Swedish University of Agricultural Sciences, S-901 83 Umeå, Sweden
On leave from: Forestry and Forest Products Research Institute, P.O. Box 16, Tsukuba Norin Kenkyu Danchi-nai, Ibaraki 305, Japan

NAMKOONG, G. : U. S. Department of Agriculture, Forest Service, Genetics Department, Box 7614, North Carolina State University, Raleigh, North Carolina 27695-7614, USA

ROSS, M.D. : Abteilung für Forstgenetik und Forstplanzenzüchtung, Universität Göttingen, Büsgenweg 2, D-3400 Göttingen, Federal Republic of Germany

RUDIN, D. : Department of Forest Genetics and Plant Physiology, The Swedish University of Agricultural Sciences, S-901 83 Umeå, Sweden

SAKAI, Kan-Ichi : 23-2 Hatoyama New Town, Hatoyama-machi, Hiki-gun, Saitama-ken, Japan 350-03

SCHNABEL, A. : Department of Botany, University of Kansas, Lawrence, Kansas 66045, USA

SCHOLZ, F. : Bundesforschungsanstalt für Forst- und Holzwirtschaft, Institut für Forstgenetik und Forstpflanzenzüchtung, Sieker Landstraße 2, D-2070 Großhansdorf 2, Federal Republic of Germany

SZMIDT, A.E. : Department of Forest Genetics and Plant Physiology, The Swedish University of Agricultural Sciences, S-901 83 Umeå, Sweden
On leave from: Institute of Dendrology, Polish Academy of Sciences, PL-62-035 Kórnik, Poland

WILLIAMS, D.A. : University of Aberdeen, Department of Forestry, St. Machar Drive, Aberdeen, United Kingdom. AB9 2UU

YAZDANI, R. : Department of Forest Genetics and Plant Physiology, The Swedish University of Agricultural Sciences, S-901 83 Umeå, Sweden

ZIEHE, M. : Abteilung für Forstgenetik und Forstpflanzenzüchtung, Universität Göttingen, Büsgenweg 2, D-3400 Göttingen, Federal Republic of Germany